Nelson Int
Mathemat

Workbook 5

Name:

2nd edition

Nelson Thornes

Contents

Contents

Beyond 999

1 **Complete the chart. Draw dots to represent the number for each place value. The first one has been done for you.**

Number	Ten thousand 10 000s	Thousands 1000s	Hundreds 100s	Tens 10s	Units 1s
12 350	•	••	•••	•••••	
58 016					
22 483					
18 095					
73 812					
82 743					
69 741					
12 694					
93 621					
21 762					

2 **Calculate in your head:**

a 43 188 + 100 =

b 43 188 + 1000 =

c 43 188 + 10 000 =

d 43 188 – 100 =

e 43 188 – 1000 =

f 43 188 – 10 000 =

g 34 347 + 200 =

h 34 347 + 2000 =

i 34 347 + 20 000 =

j 34 347 – 200 =

k 34 347 – 2000 =

l 34 347 – 20 000 =

m 20 000 + 100 =

n 30 000 + 1000 =

o 50 000 + 10 000 =

p 20 000 – 100 =

q 30 000 – 1000 =

r 50 000 – 10 000 =

see Student Book page 6

What is it worth?

12$\underline{5}$609

The digit 5 is worth five thousand because of its position in the number.

H.Th	T.Th	Th	H	T	U
1	2	5	6	0	9

Write in words what each underlined digit is worth.

1 125 $\underline{6}$09

2 125 60$\underline{9}$

3 58 4$\underline{6}$3

4 58 $\underline{4}$63

5 $\underline{5}$8 463

6 7$\underline{5}$2 186

7 752 1$\underline{8}$6

8 $\underline{7}$52 186

9 8$\underline{3}$7 612

10 $\underline{6}$27 141

11 766 $\underline{4}$31

12 $\underline{1}$79 487

see *Student Book page 6*

Rounding to 10 or 100

1 Circle the numbers that round to 250 when they are rounded to the nearest ten.

244	256	259	241
253	252	249	248
245	258		

2 Circle the numbers that round to 1500 to the nearest 100.

1569	1467	1416	1456
1575	1483	1590	1532
1439	1522		

3 Round each number to the nearest 10 and to the nearest 100.

Number	To the nearest 10	To the nearest 100
369		
481		
1402		
8492		
6445		
2569		
1385		
8884		
5495		
7783		

4 **A newspaper reported that 1300 people attended a cricket match.**

a If this number was rounded to the nearest 10, what is the smallest and greatest number of people who could have attended?

b If this number was rounded to the nearest 100, what is the smallest and greatest number of people who could have attended.

c Tell your partner how you worked out your answers.

see Student Book page 9

Rounding to 1000s

Round each number to its nearest thousand.

Join it to its nearest thousand with a line.

The first one has been done for you.

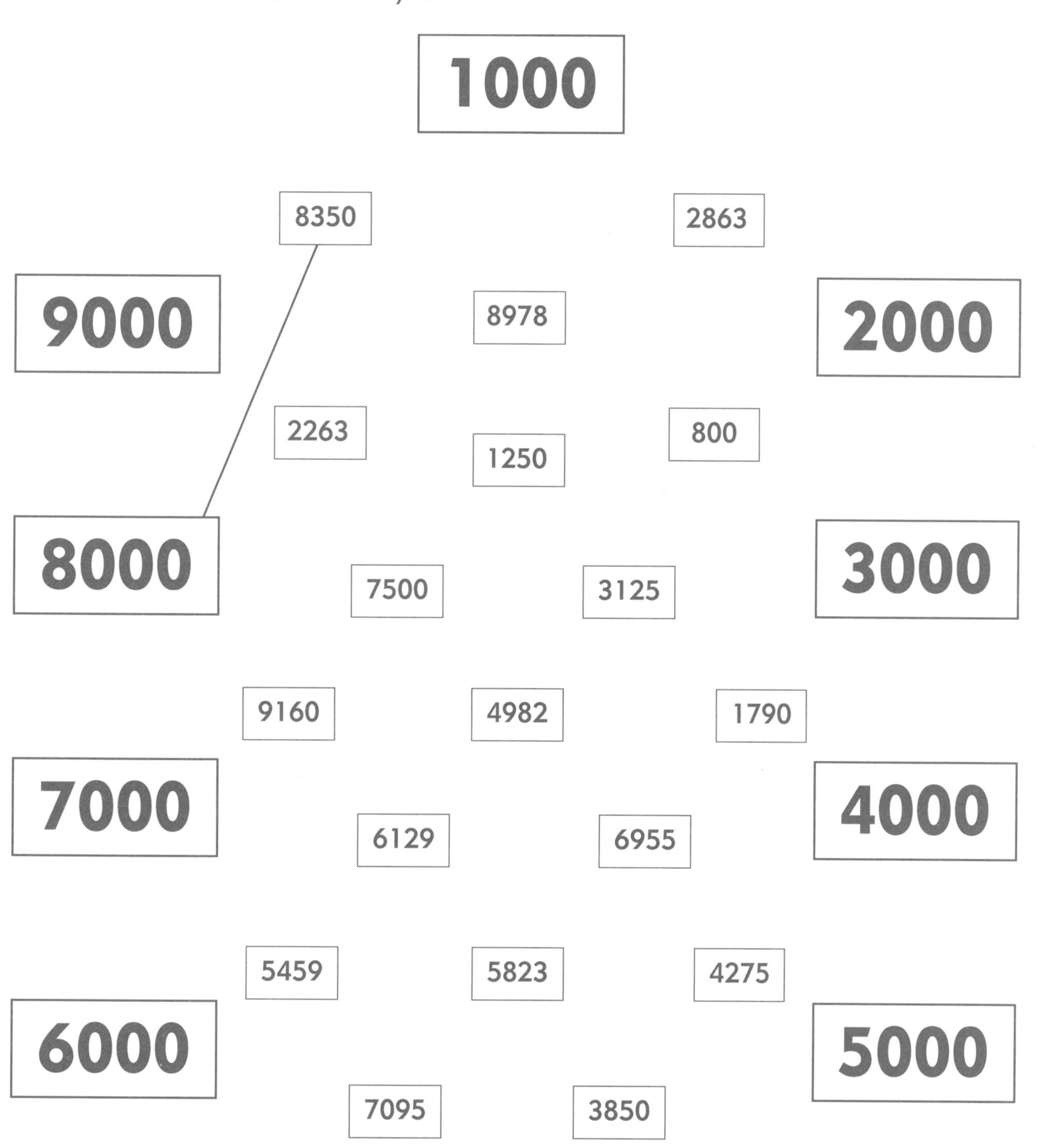

see Student Book page 10

Using inverse operations

For each number sentence, check whether it is true or false. Write the inverse number operation to check. Then write true or false. The first one has been done for you.

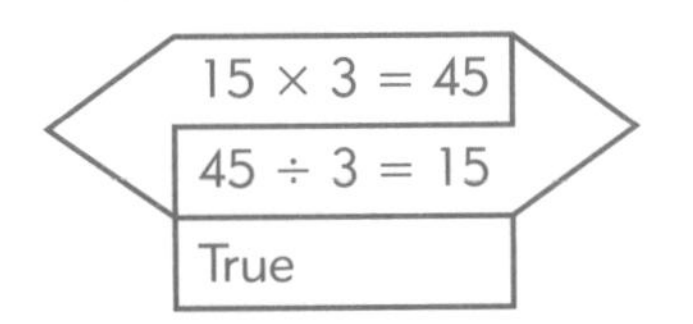

multiply ← inverse → divide

add ← inverse → subtract

You can use a calculator to help you.

1

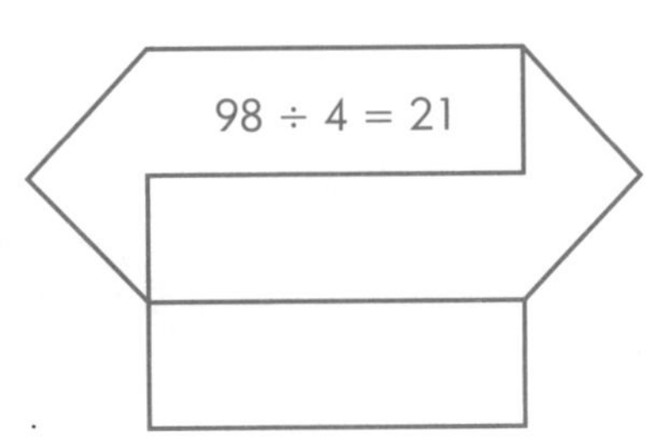

2

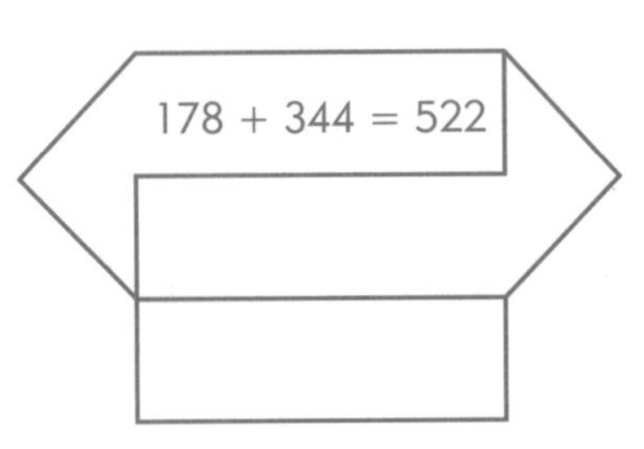

3

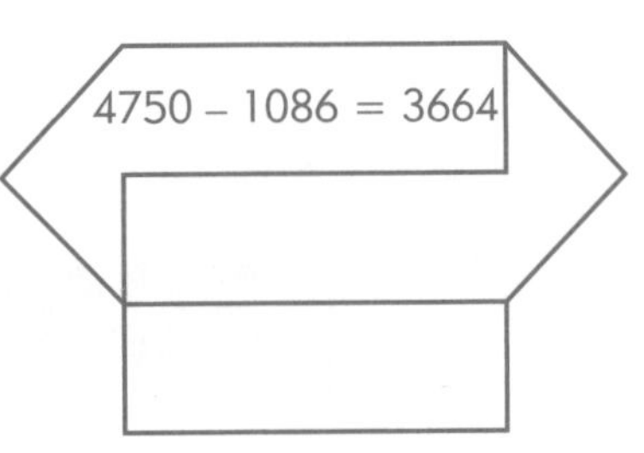

4

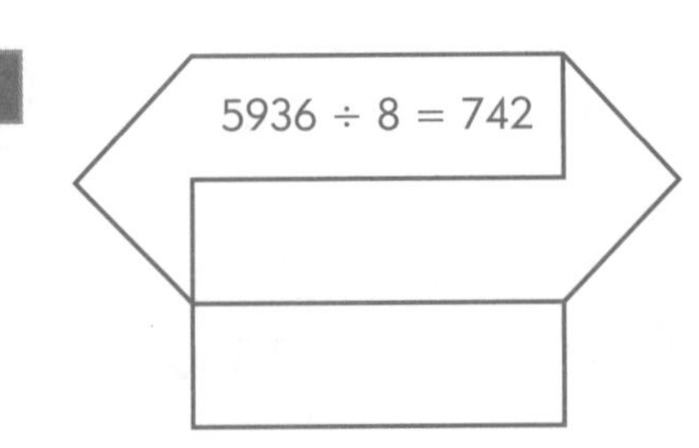

5

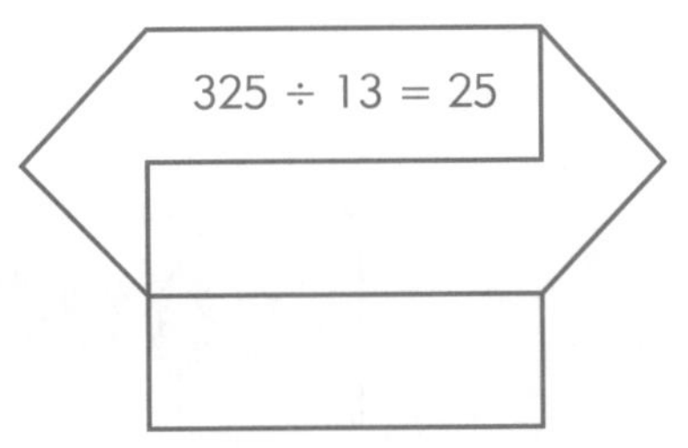

6

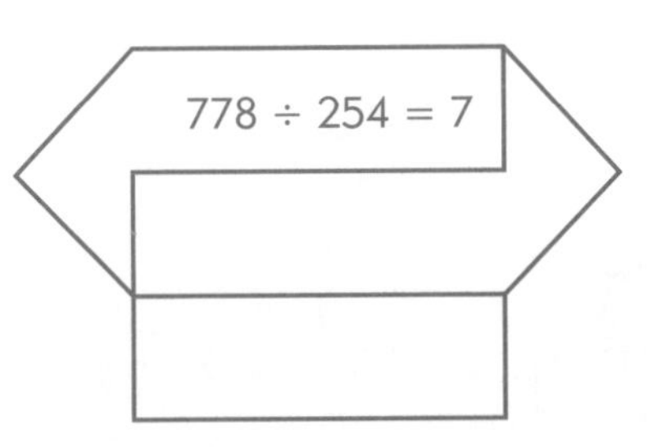

7

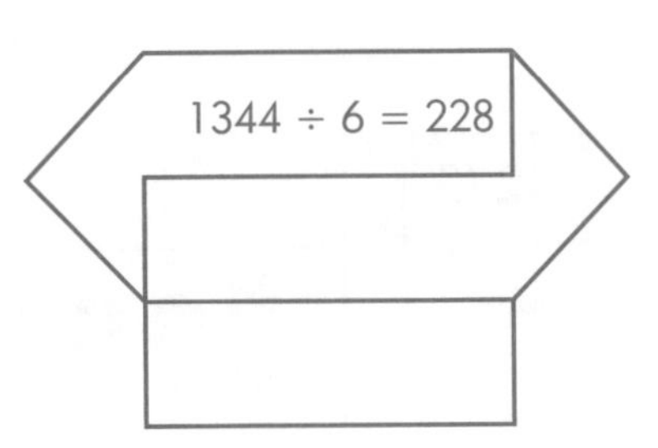

8

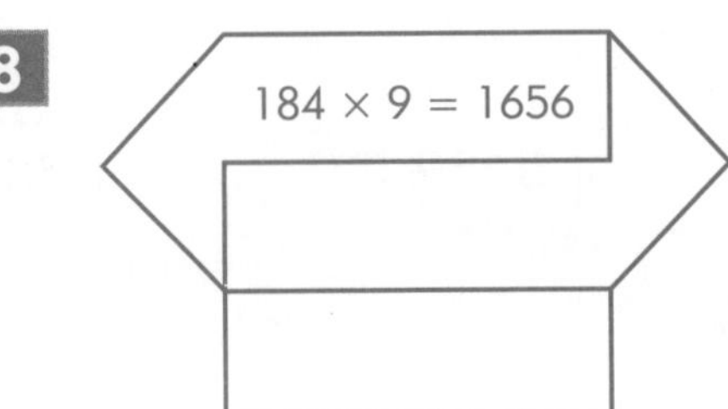

see Student Book page 12

Order of operations

1 For each calculation:
- circle the part you would do first
- underline the bit you would do second
- work out the answers.

a $(10 - 2) - 3$

b $8 + 5 - 3$

c $(12 - 9) \times 24 - 6$

d $144 - 21 \times 2$

e $10 + 7 - 2 \times 6$

f $36 \div 6 - 2$

g $5 - 10 \div 2$

h $15 \div 3 - (3 + 2)$

i $3 \times 4 - 2 \times 6$

j $7 - 24 \div 6$

k $6 \times 3 - 17$

l $3 \times 3 - 3$

m $4 \times 2 - 16 \div 2$

n $2 + 2 + 2 \times 2$

o $3 \times 9 - 5 \times 5$

p $(8 + 3) \times (20 \div 2) \div 11$

q $2 \times (6 - 3) + 5$

r $(12 + 6) \div (5 - 2)$

s $(99 + 22) \div 11$

t $8 \div (16 - 14) - 1$

---

2 Make up six calculations of your own. Make sure there are at least three steps in each one.

a ____________ **b** ____________ **c** ____________

d ____________ **e** ____________ **f** ____________

3 Swap with a partner. Work out the answers to your partner's calculations.

see Student Book page 13

Missing operation signs

1 **Put in the operation signs (+, –, ÷, ×) to make these number sentences true. You may also need to use brackets.**

a $2 + 3 \square 4 = 1$

b $2 \square 3 - 4 = 2$

c $2 \square 3 \square 4 = 3$

d $2 \square 3 \square 4 = 10$

e $8 \square 3 \square 2 = 22$

f $8 \square 3 \square 2 = 40$

g $8 \square 3 \square 2 = 12$

h $8 \square 3 \square 2 = 2$

i $8 \square 3 \square 2 = 10$

j $8 \square 3 \square 2 = 9$

---

2 **Put in the brackets to make these statements true.**

a $5 \times 7 - 3 = 20$

b $28 - 13 - 6 = 21$

c $6 - 5 \times 12 = 12$

d $38 - 23 + 17 - 12 = 10$

e $48 \div 12 + 6 \times 6 = 1$

f $23 - 18 - 5 \times 7 = 0$

g $5 \times 6 \div 13 - 10 = 10$

h $8 \times 7 - 2 = 40$

---

3 **Use <, > or = to make true statements.**

a $(8 + 7) \times 5$ ______ $8 + 7 \times 5$

b $7 + 2 \times 4$ ______ $5 \times 2 + 6$

c $5 \times 9 \div 3$ ______ $8 + 4 \times 8$

d $23 - 12 \div 6$ ______ $2 + 6 \times 3$

e $5 \times 5 \div 5$ ______ $5 \div 5 \times 5$

f $36 \div 12 \div 3$ ______ $24 \div 12 \div 2$

Remember:

> means 'greater than'

< means 'less than'

= means 'equal to'

see Student Book page 13

Working in order

1 The answers to these calculations are correct, but the brackets have been left out of the problems. Add brackets to make each number sentence true. Some number sentences might not need brackets.

a 9 × 2 + 3 = 45

b 16 – 7 × 3 = 27

c 20 ÷ 4 + 1 = 4

d 20 ÷ 4 + 1 = 6

e 18 + 9 ÷ 3 = 9

f 64 ÷ 8 – 6 = 2

g 6 + 6 × 3 = 36

h 10 – 4 × 5 = 30

i 5 + 2 × 3 + 7 = 25

j 12 + 6 ÷ 7 – 4 = 14

k 7 + 10 – 5 ÷ 2 = 6

l 6 – 3 × 2 = 6

m 10 × 6 + 4 = 100

n 27 – 14 – 8 = 21

---

2 Write whether each statement is true or false.

a (100 + 10) + 2 = 100 + (10 + 2) ____________

b (100 x 10) × 2 = 100 x (10 x 2) ____________

c (100 – 10) – 2 = 100 – (10 – 2) ____________

d (100 ÷ 10) ÷ 2 = 100 ÷ (10 ÷2) ____________

e 80 – 5 – 5 – 5 – 5 – 5 – 5 = 80 – (5 + 5 + 5 + 5 + 5 + 5) ____________

f 64 ÷ 2 ÷ 2 ÷ 2 ÷ 2 ÷ 2 ÷ 2 = 64 ÷ (2 x 2 x 2 x 2 x 2 x 2) ____________

---

3 Draw a flow diagram (on a separate piece of paper) to teach someone how to work in the correct order in maths.

see Student Book page 14

Are they the same?

1 **Here are some pairs of expressions.**

Write whether each pair gives the same or a different result.

a $2 + 3 \times 4$ and $3 \times 4 + 2$ ____________

b $2 \times 12 \div 3$ and $12 \div 3 \times 2$ ____________

c $50 - 25 - 5$ and $25 - 5 - 50$ ____________

d $25 \times 4 - 8$ and $8 - 25 \times 4$ ____________

e $25 \times 4 \div 5$ and $5 \div 25 \times 4$ ____________

---

2 **Fill in the operations and brackets to make true statements.**

a 7 3 5 4 = 46

b 7 3 5 4 = 22

c 7 3 5 4 = 24

d 7 3 5 4 = 80

e 7 3 5 4 = 200

---

3 **Choose four digits of your own. Use brackets and operations as in question 2. Use the space below to make as many different numbers as you can. Choose one of the numbers you have made. Can you make it in more than one way using the same four digits?**

see Student Book page 14

Finish the shapes

These shapes are only half-drawn.
Use the spots to complete the shapes.
Write what shape you have drawn on
the line under each shape.

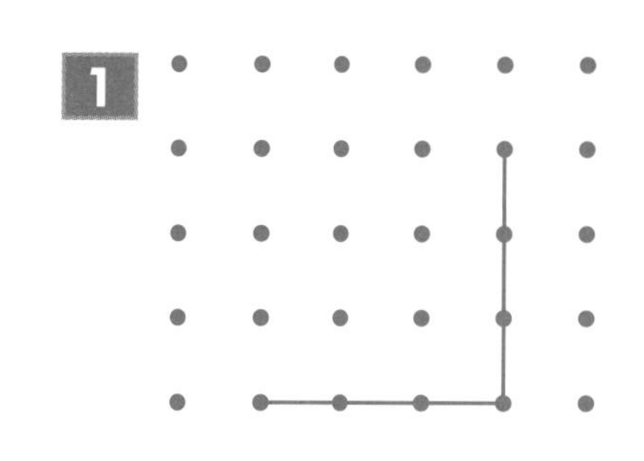

2

3

4

5

6

7

8

see Student Book page 16

Symmetry in polygons

Write the name of each polygon.
Draw at least one line of symmetry on each.

1

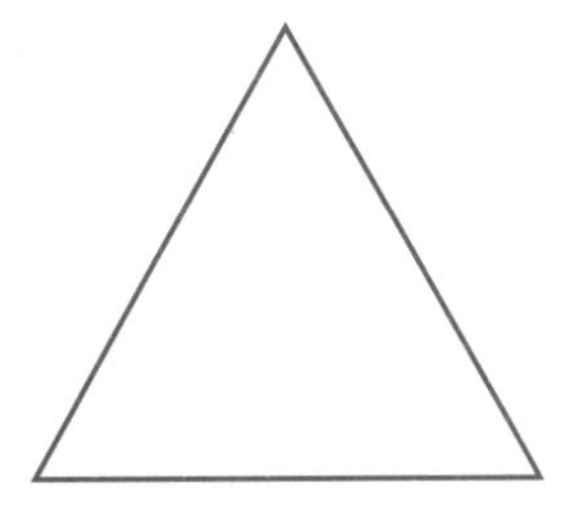

____

2

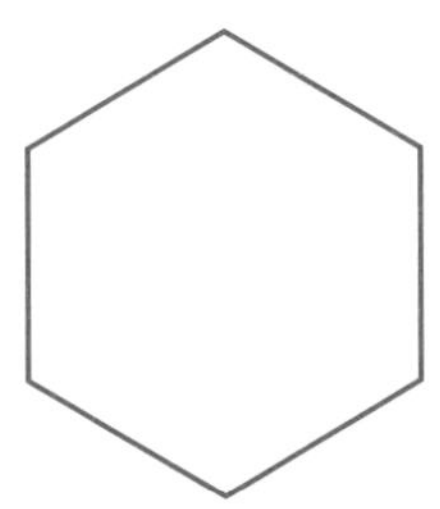

____

3

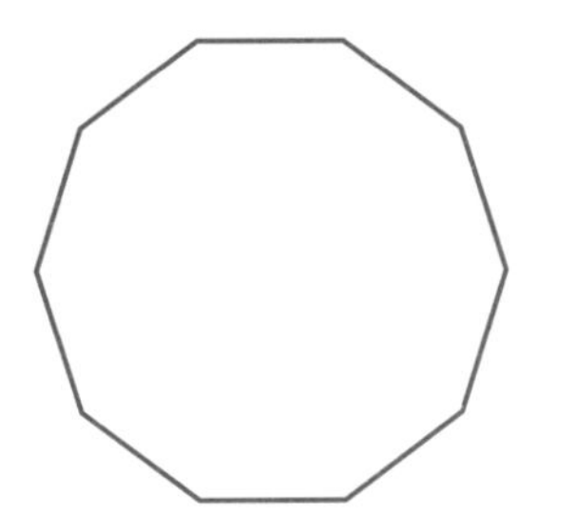

____

4

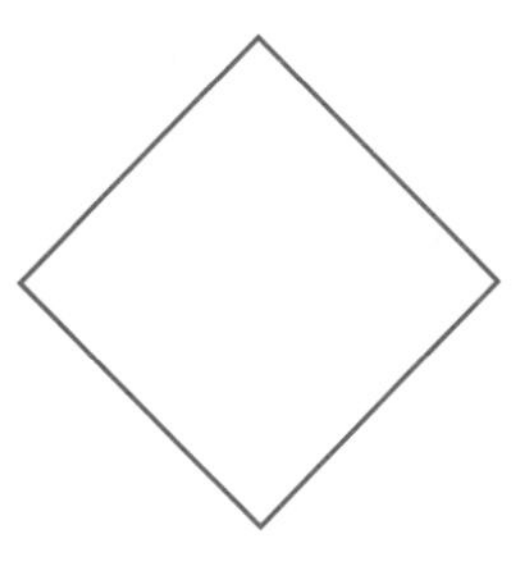

____

5

____

6

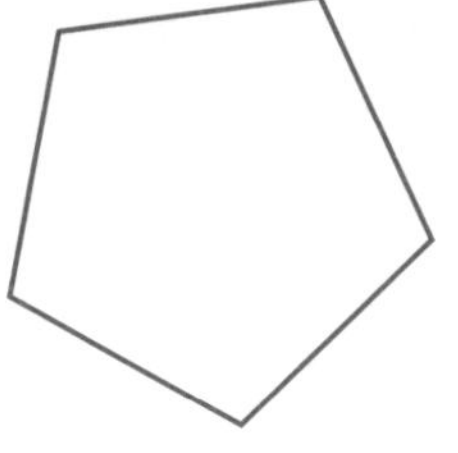

____

7

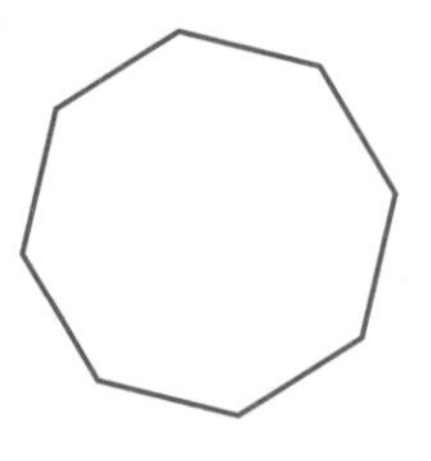

____

8

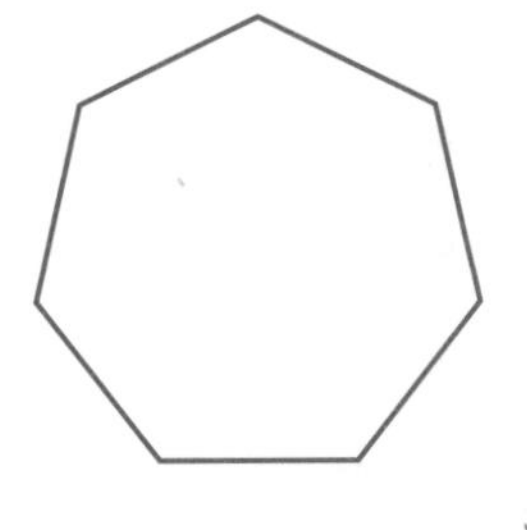

____

see Student Book page 17

More symmetry

On these shapes, draw any lines of symmetry in different colours.

You can check with a mirror. If your line of symmetry is correct, you will see the other half of the shape in the mirror.

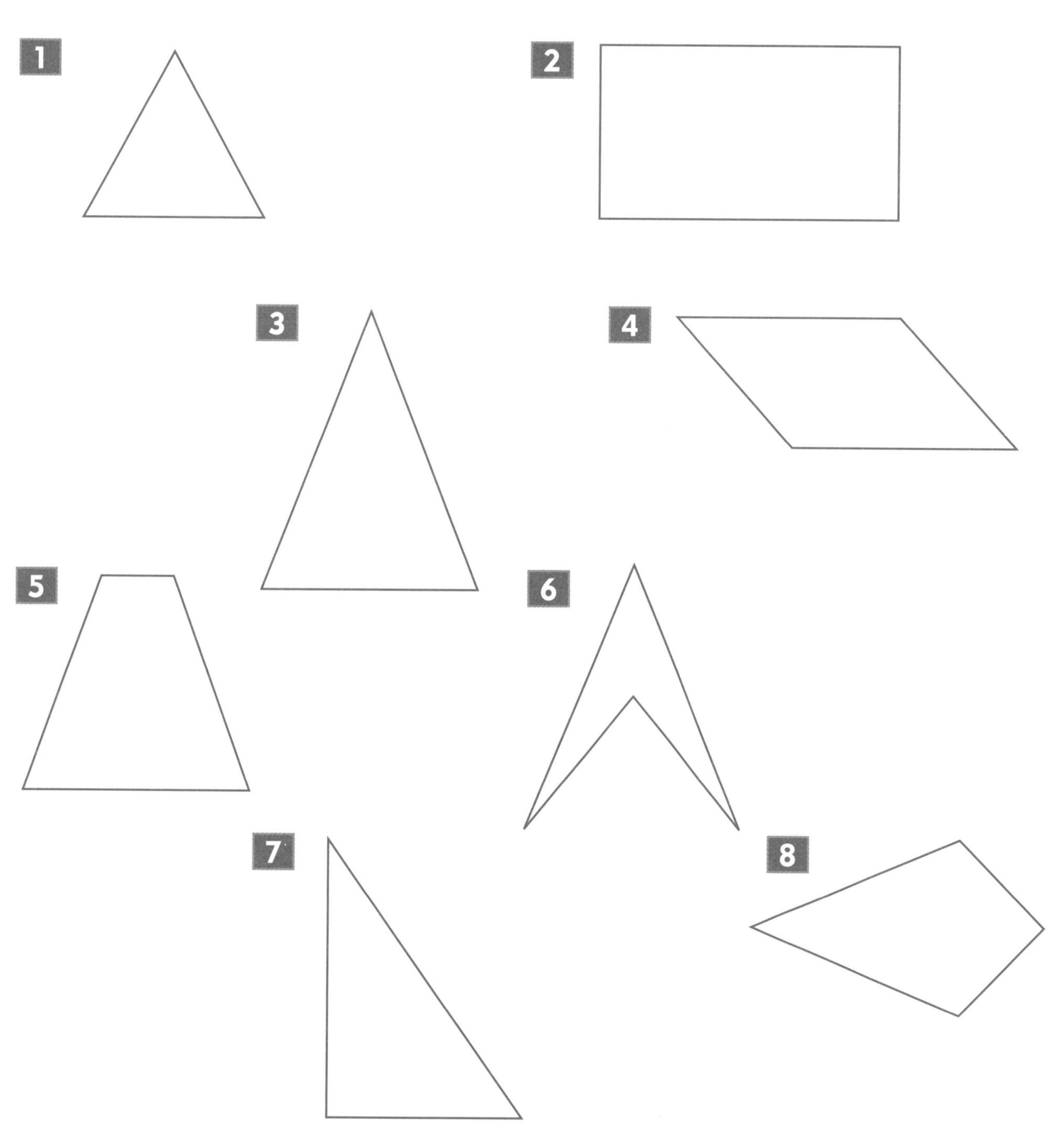

see Student Book page 17

Symmetry patterns

Complete each pattern so it is symmetrical about both lines of symmetry.

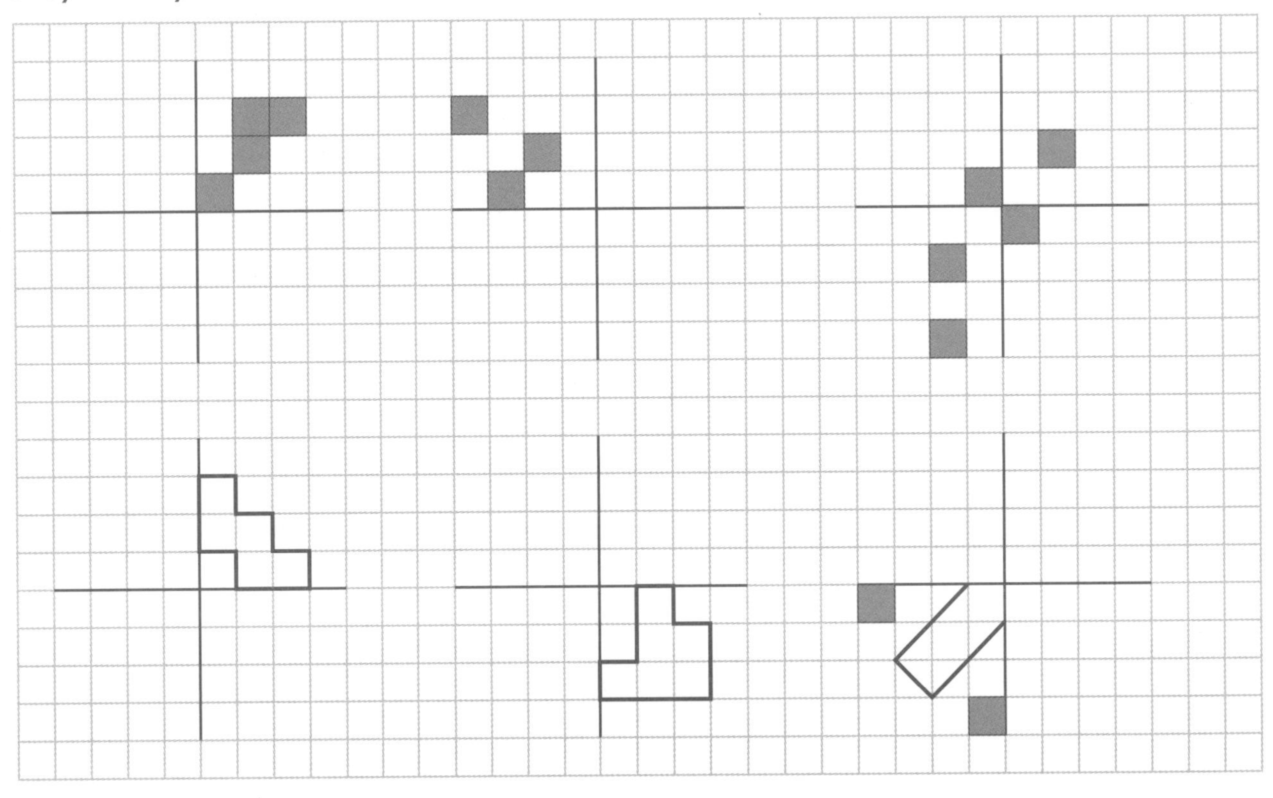

Draw your own symmetrical patterns using these lines of symmetry.

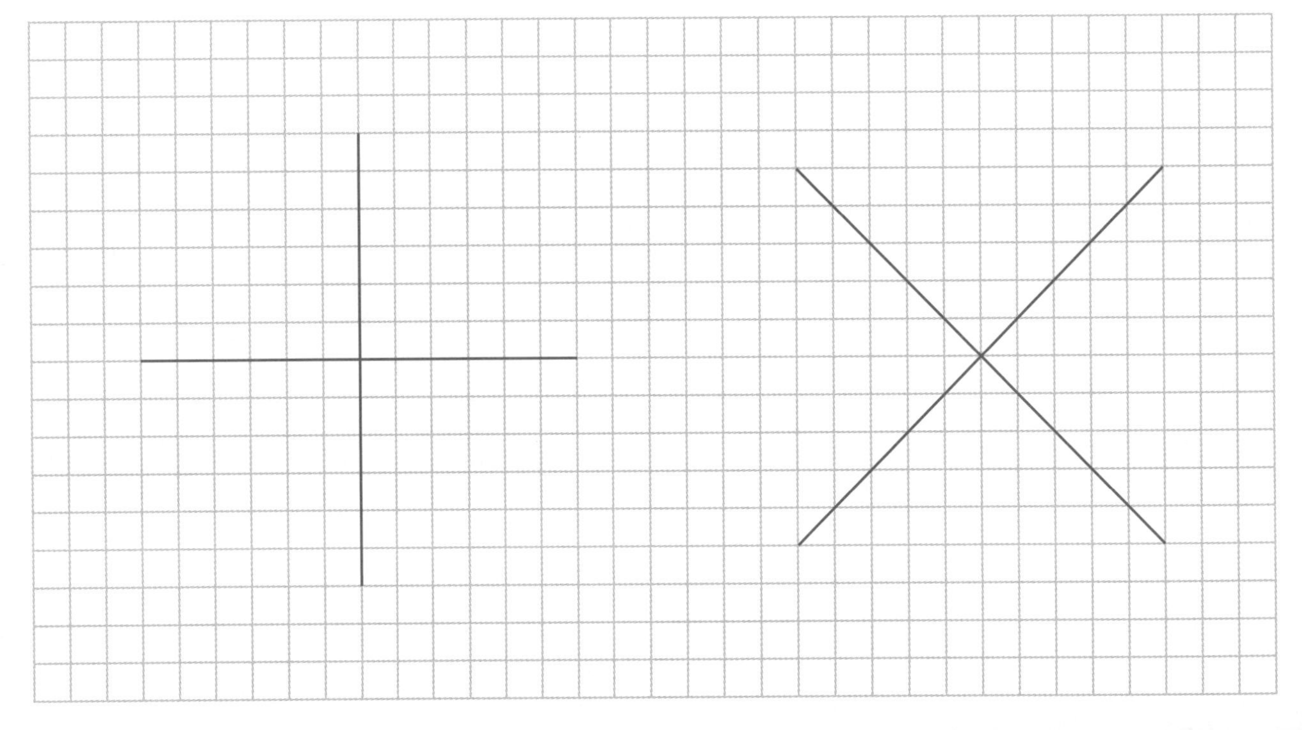

see Student Book page 18

Spinning logos

Many common shapes and patterns have rotational symmetry.

1 **What order of symmetry do these have?**

a

b

c

d

---

2 **Design a logo for your class. Decide what order of rotational symmetry it should have. Draw it here.**

see Student Book page 19

National flags

The flag of Jamaica has two lines of symmetry. If it was lifted and rotated about its centre point, it would fit into its own outline in two different positions. So it has rotational symmetry of order 2.

For each flag below, write the number of lines of symmetry and order of rotational symmetry.

1

Australia

lines ______________

order ______________

2

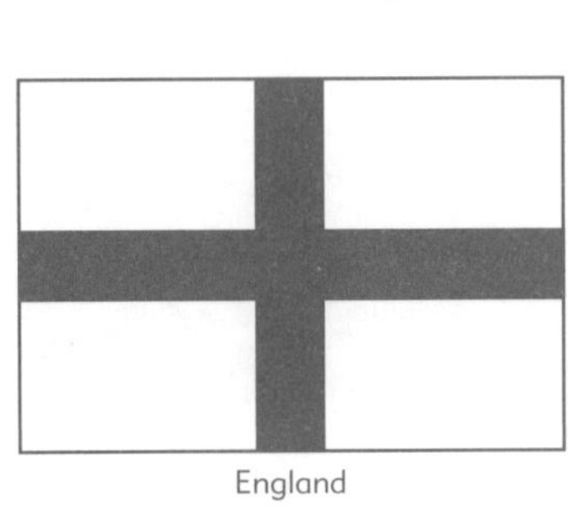

England

lines ______________

order ______________

3

France

lines ______________

order ______________

4

India

lines ______________

order ______________

5

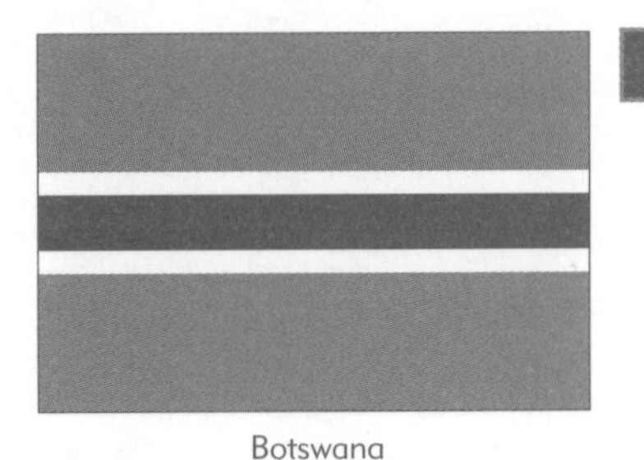

Botswana

lines ______________

order ______________

6

Canada

lines ______________

order ______________

7

Trinidad and Tobago

lines ______________

order ______________

8

Thailand

lines ______________

order ______________

9

Switzerland

lines ______________

order ______________

see Student Book page 19

Classifying triangles

1 Colour the odd triangle out in each set. You may need to measure the sides and angles to help you decide which one is the odd one out.

2 Write a sentence saying why the coloured one does not fit with the others in the set.

see *Student Book page 22*

Showing time on different clocks

Complete the chart to give the times in words, as a.m. and p.m., and in 24-hour notation. Remember that 24-hour time always uses four digits.

The first three have been done for you.

In words		
half past eight in the morning		08:30
three o'clock in the afternoon		15:00
quarter to eight at night		19:45
seven o'clock in the morning		:
half past ten at night		:
quarter past four in the afternoon		:
quarter to one in the afternoon		:
twenty-five to five in the afternoon		:
five to eleven at night		:

see Student Book page 23

Holiday activities

On a holiday camp, there is a very full programme of activities to choose from. These are the activities on offer in the morning:

Start time	Activity	Length of time in hours
6.00 a.m.	Hiking	3
6.35 a.m.	Surfing	2
8.30 a.m.	Riding	$1\frac{1}{2}$
8.45 a.m.	Target shooting	1
9.30 a.m.	Painting	$2\frac{1}{2}$
10 a.m.	Judo	$1\frac{1}{2}$
10 a.m.	Archery	$1\frac{1}{2}$
12 noon	Lunch	1

1 **Choose five different activities from the list. Write the start and finish time for each one. Use 24-hour time.**

2 **Write down how many of the five on your list you would be able to do in one day. Explain how you worked out your answer.**

see *Student Book page 23*

Units of time

1 **There are seven units of time hidden in this word search.**

a Find the units of time.

b Write them in order from the shortest to the longest.

H	O	U	R	Q	Y	L
S	E	L	W	E	E	K
B	P	F	X	D	A	Y
A	J	Y	H	C	R	N
M	I	N	U	T	E	G
M	O	N	T	H	R	W
T	S	E	C	O	N	D

______

______

______

______

______

______

______

2 **Complete.**

a There are __ days in 1 week.

b 8 weeks = __ days

c 35 weeks = __ days

d There are __ weeks in a year

e 84 days = __ weeks

f 91 days = __ weeks

g 175 days = __ weeks

h 16 weeks is about __ months

i 8 months is about __ weeks

j 1 year = __ months

k 5 years = __ months

l 20 years = __ months

m 36 months = __ years

n 84 months = __ years

o 66 months = __ years

3 **How many days in?**

a 1 week 6 days

b 3 weeks

c this year plus next year

d the first six months of the year

e the last six months of the year

see Student Book page 28

Making decimals

Shade each shape to show the decimal fraction.

Write a decimal to show what fraction of the shape is unshaded.

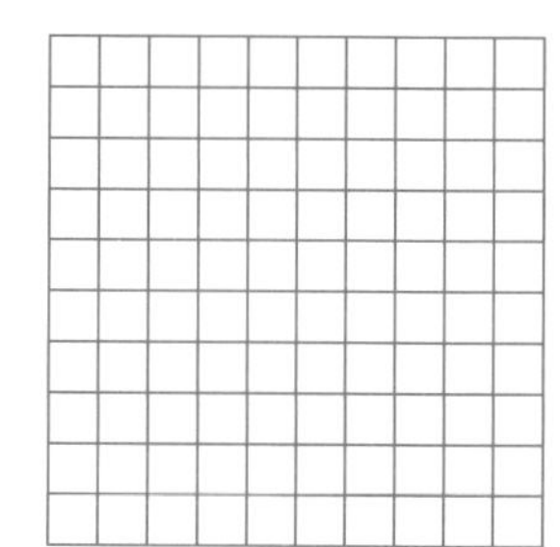

Shaded 0.4
Unshaded _____

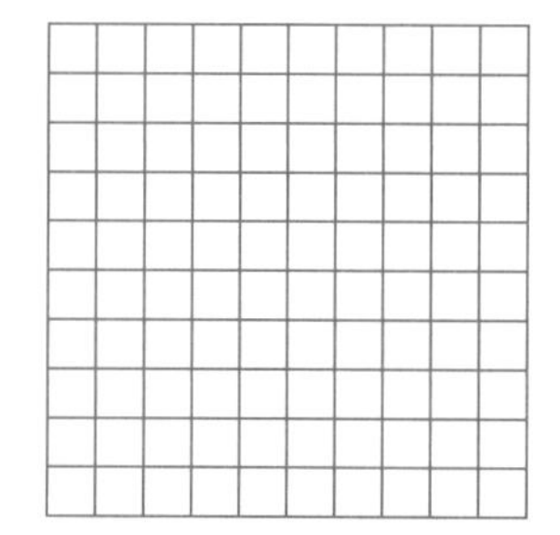

Shaded 0.43
Unshaded _____

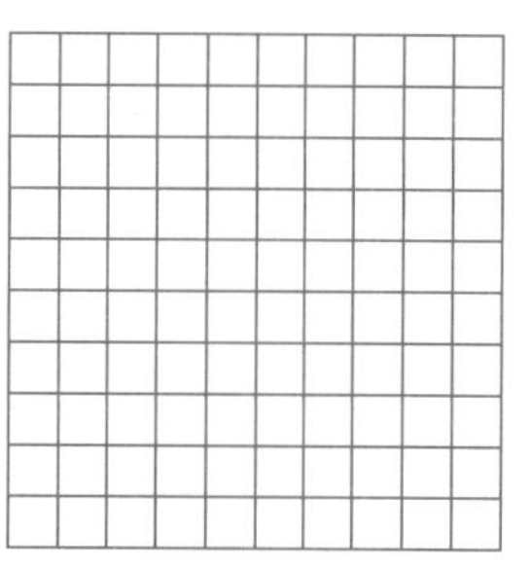

Shaded 0.9
Unshaded _____

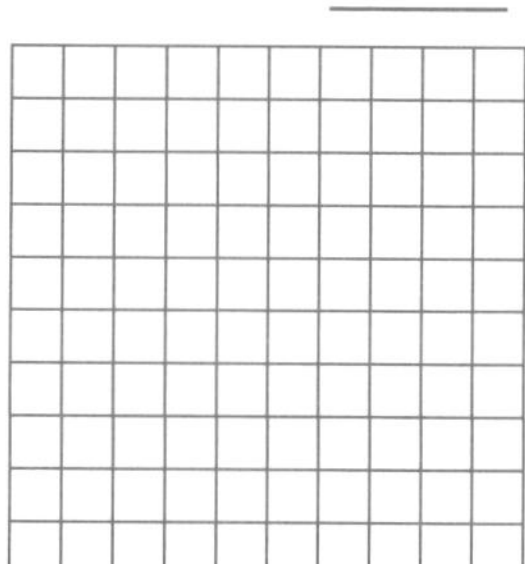

Shaded 0.5
Unshaded _____

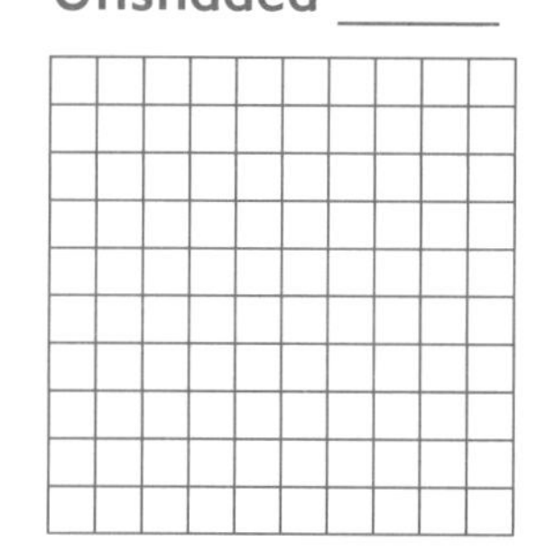

Shaded 0.05
Unshaded _____

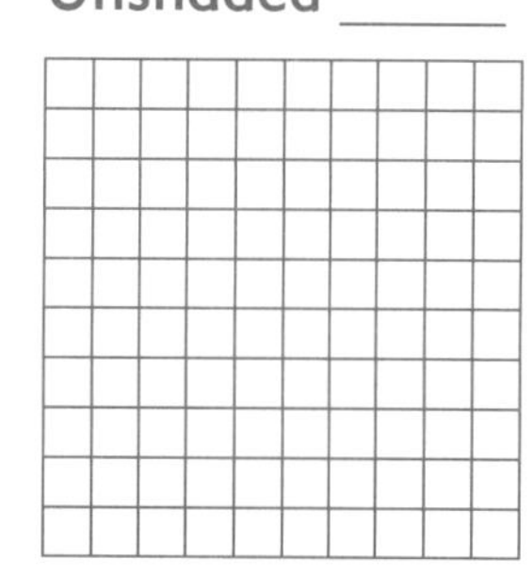

Shaded 0.55
Unshaded _____

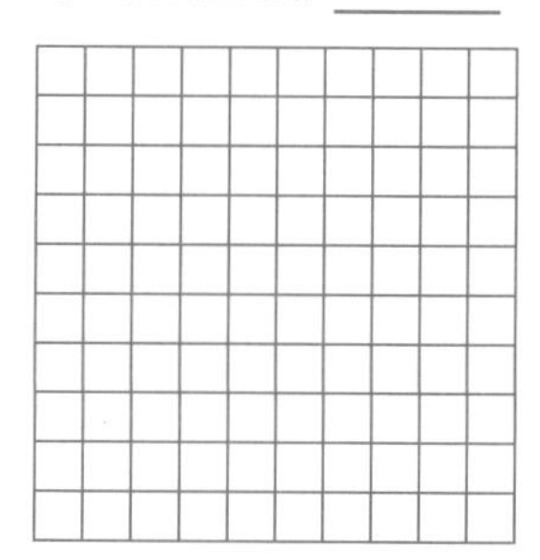

Shaded 0.27
Unshaded _____

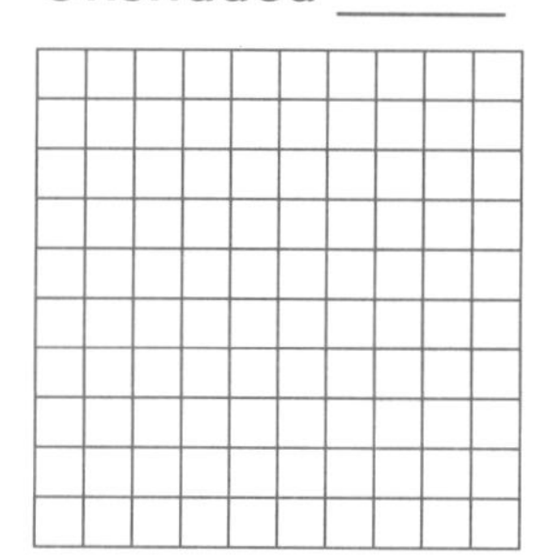

Shaded 0.72
Unshaded _____

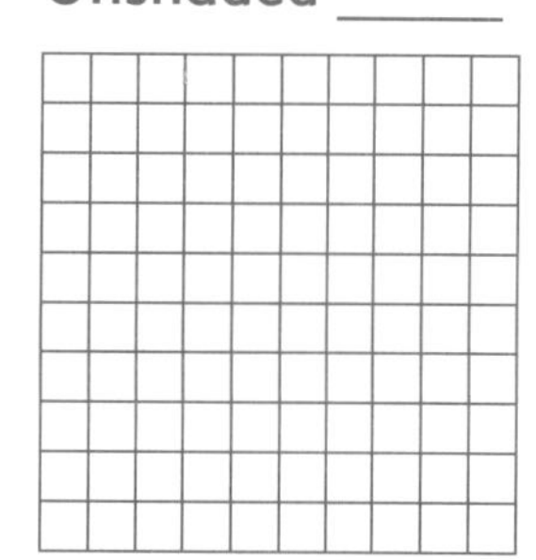

Shaded 0.07
Unshaded _____

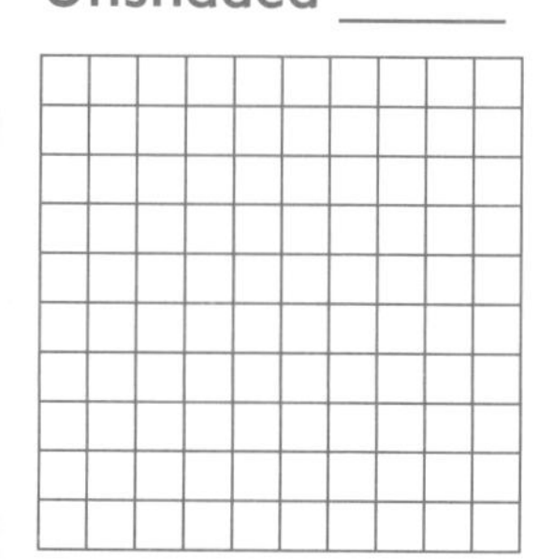

Shaded 0.9
Unshaded _____

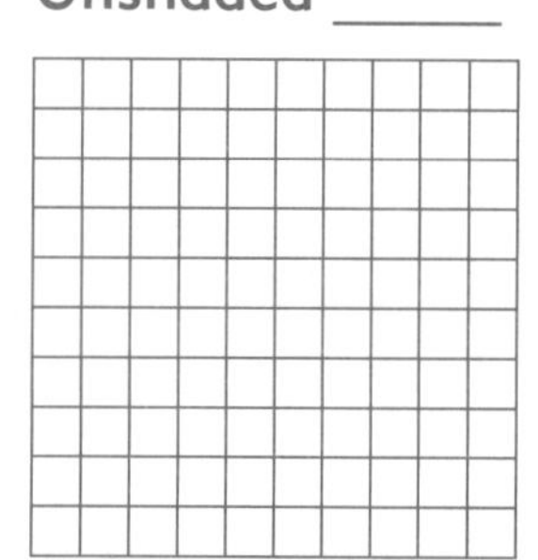

Shaded 0.09
Unshaded _____

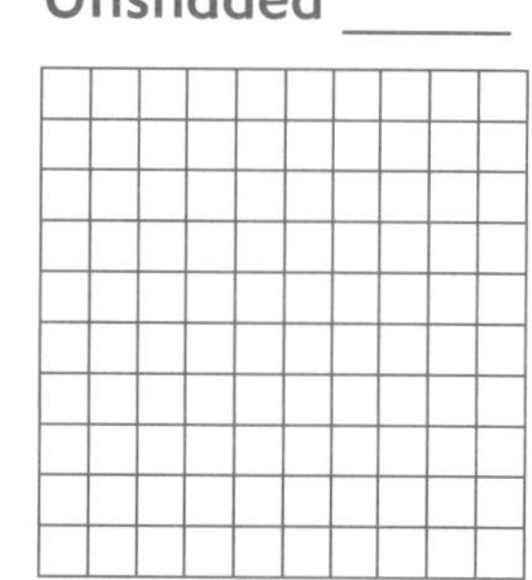

Shaded 0.99
Unshaded _____

see Student Book page 29

Locating decimals on a number line

Mark and label each set of decimals on the given number lines.

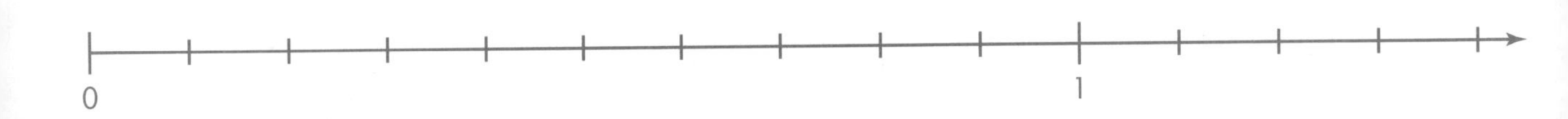

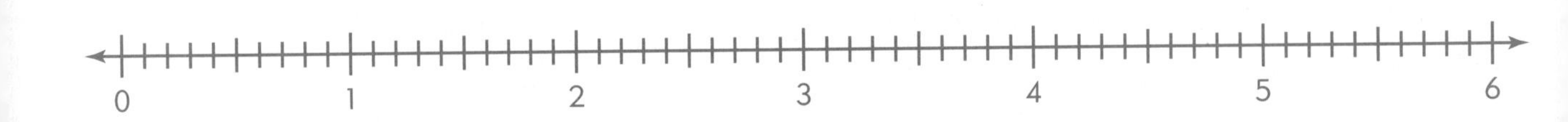

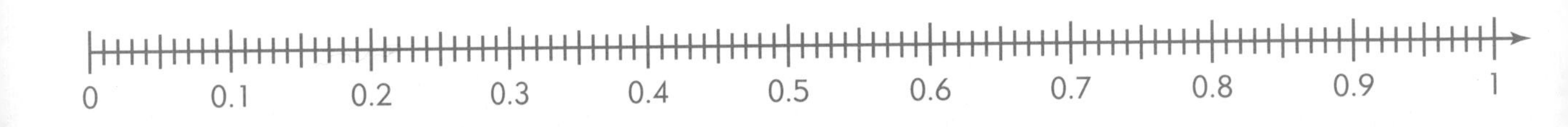

1 0.8 0.5 1.1 0.2 0.7 0.1

2 0.1 1.0 1.5 2.3 4.4 5.8

3 0.55 0.09 0.03 0.33 0.75 0.99

4 21.35 21.39 21.51 21.64 21.69 21.72

5 Fill in < or > between each pair of decimals. Use your number lines to help you decide.

a 0.3 ☐ 1 **b** 0.05 ☐ 0.5 **c** 2.5 ☐ 2.9

d 0.7 ☐ 0.69 **e** 0.09 ☐ 0.1 **f** 0.33 ☐ 0.3

see Student Book page 30

Units of length

Complete the tables by filling in the equivalent measurements.

Kilometres	Metres
1	
5	
7.5	
12	
135	

Metres	Centimetres
1	
9	
28	
98	
150	

Centimetres	Millimetres
1	
7	
15	
67	
189	

Metres	Millimetres
1	
8	
15	
27	
112	

see Student Book page 34

Measure in different units

Measure each item and write the measurement in two different ways.

Item	In millimetres	In centimetres
the length of my thumbnail		
span of my left hand		
length of my shoe		
length of a pair of scissors		
length of a pencil		
width of this book		
height of a coffee mug		
length of a stapler		
width across a pair of spectacles		
length of a belt		

see Student Book page 34

Measuring scales

1 **The arrows represent measurements on each scale. Estimate what each measurement is.**

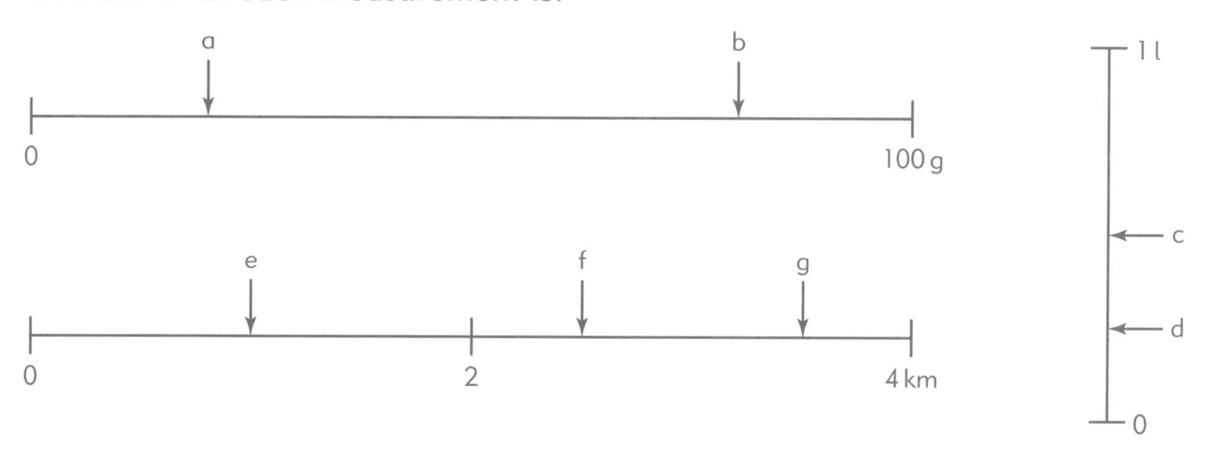

2 **Draw more liquid in each jug to show the given measurements.**

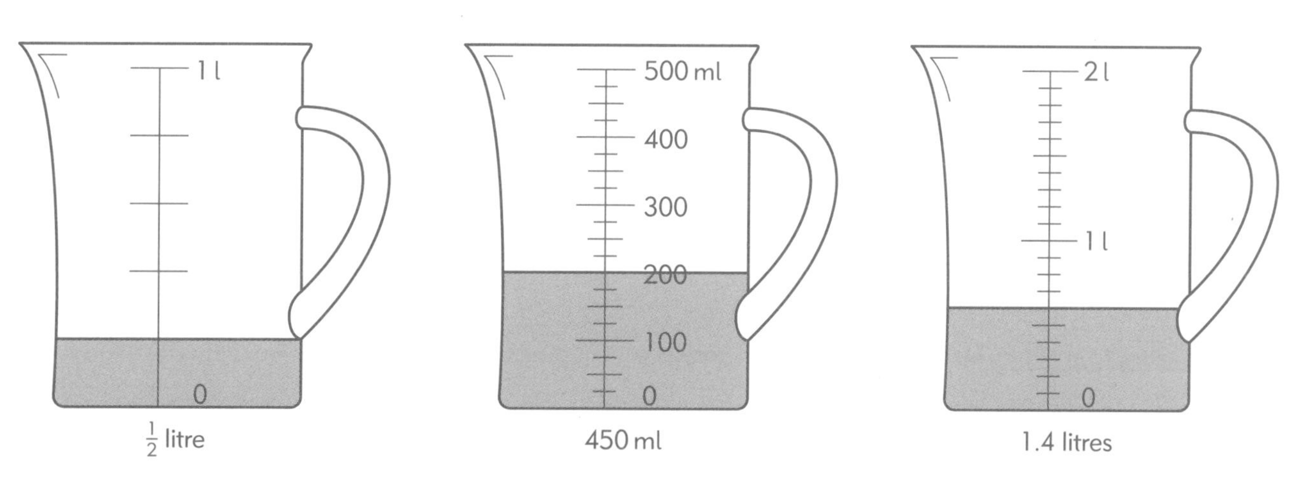

$\frac{1}{2}$ litre

450 ml

1.4 litres

3 **Draw arrows on each scale to show the given measurements.**

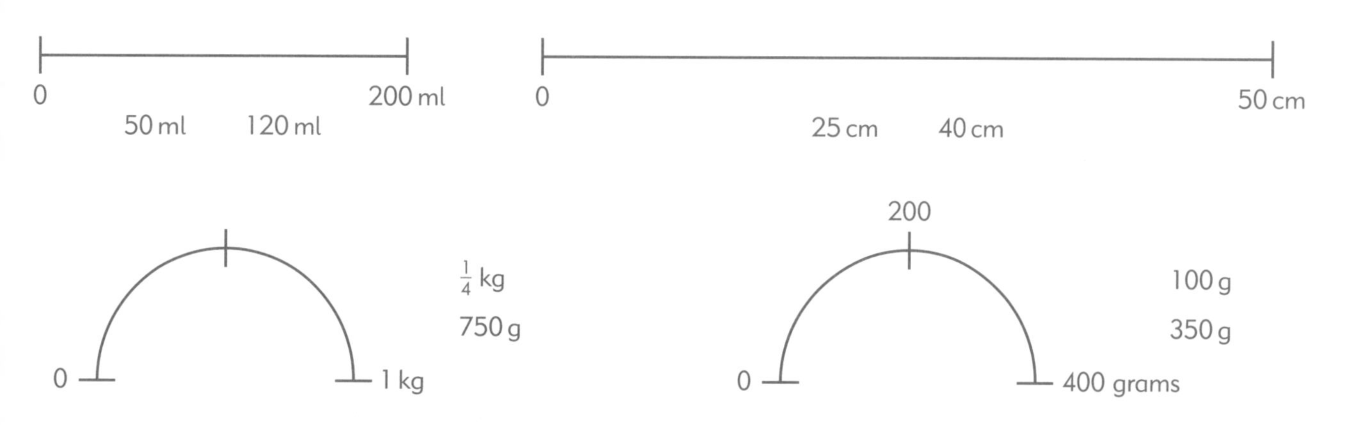

see *Student Book page 38*

Comparing scales

Draw arrows to show where the given amount would show on each scale

450g

1.2l

3.4m

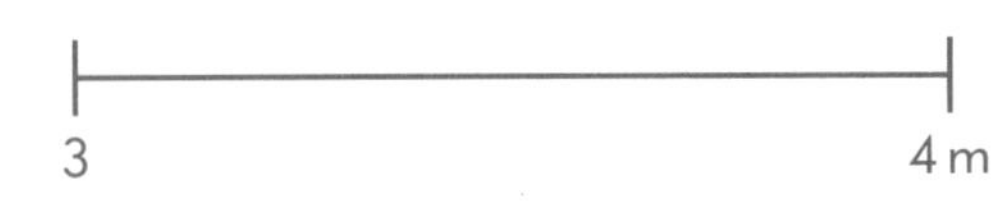

2.4kg

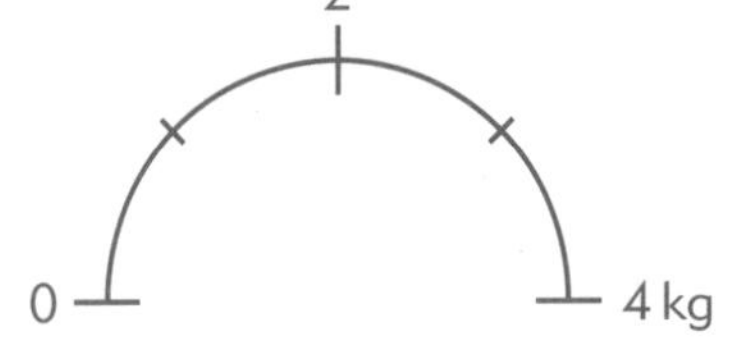

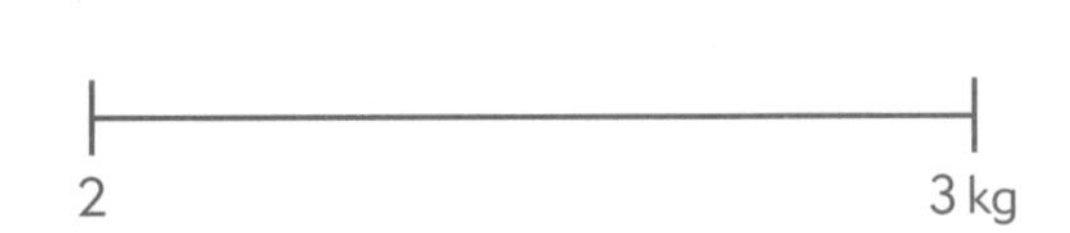

45ml

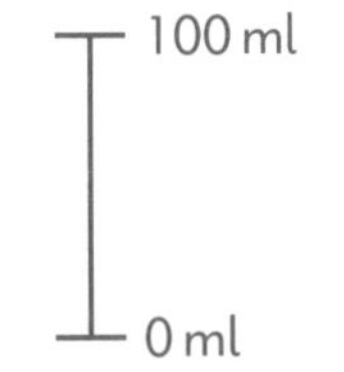

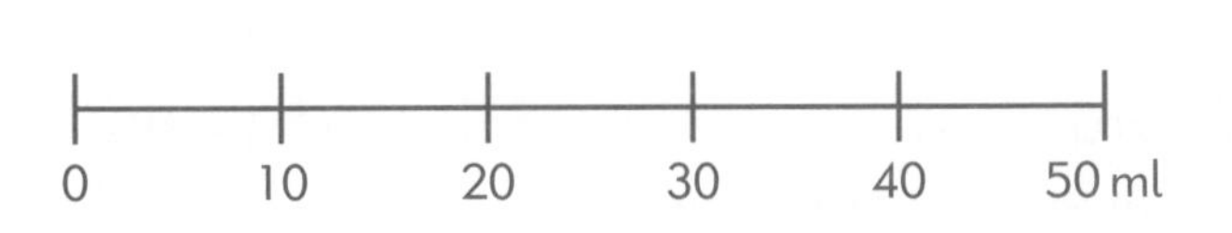

32°C

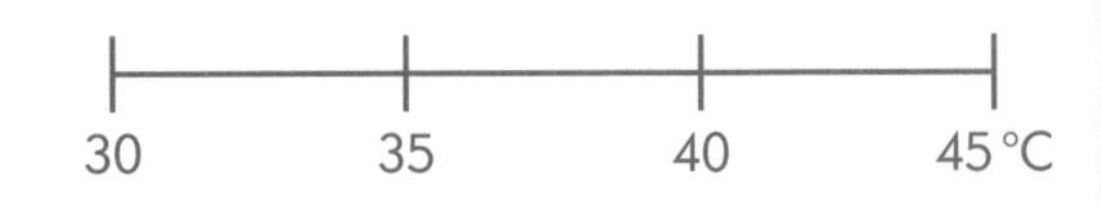

–5°C

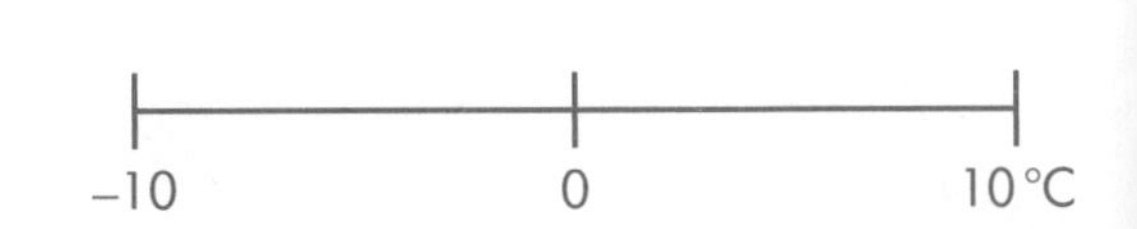

see Student Book page 38

Counting in steps

Count on or back in the given steps to complete each set of numbers.

When you have finished, use a calculator to check your answers.

Count on in 12s	56						

Count back in 12s	36						

Count on in 11s	20						

Count back in 11s	88						

Count back in 11s	86						

Count on in 15s	205						

Count back in 15s	60						

Pete counts in fives from 132 to 500. Try to answer these questions without actually counting.

Would he count 182?

Would he count 387?

Would he count 198?

Explain how you answered these questions.

see *Student Book page 41*

Adding and subtracting by counting in groups

Fill in the missing numbers on each number line.

Write the answers to each calculation.

247 + 318 = ☐

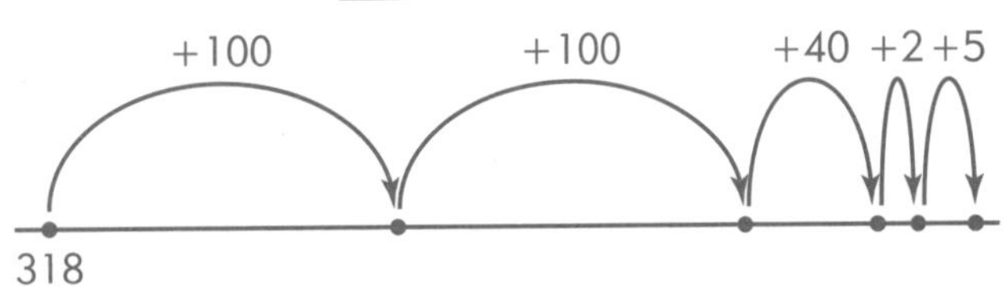

142 + 612 = ☐

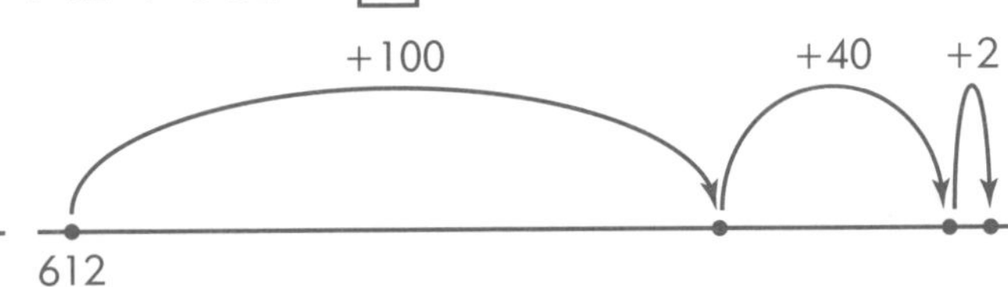

762 + 189 = ☐

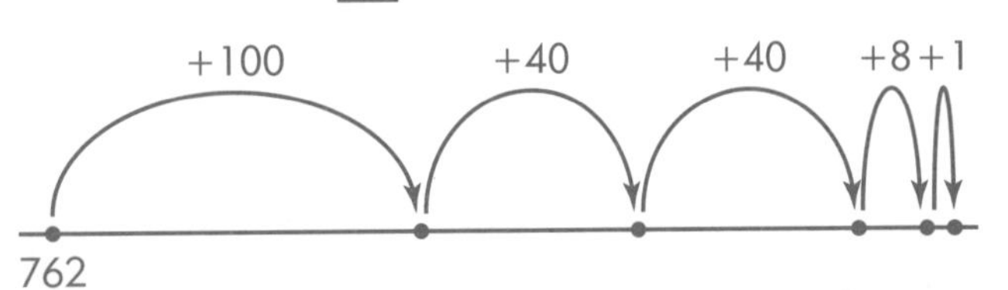

876 + 1245 = ☐

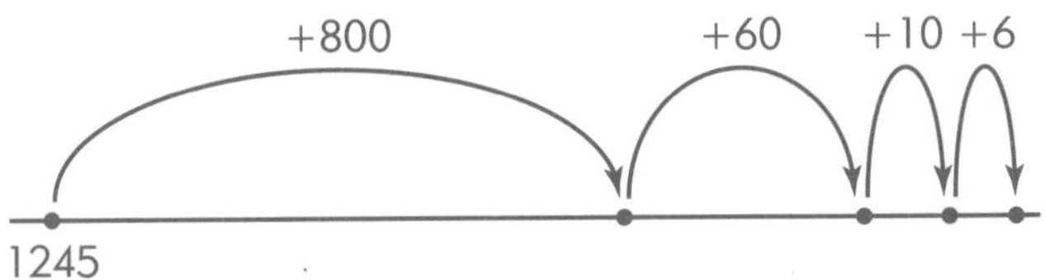

3109 + 1098 = ☐

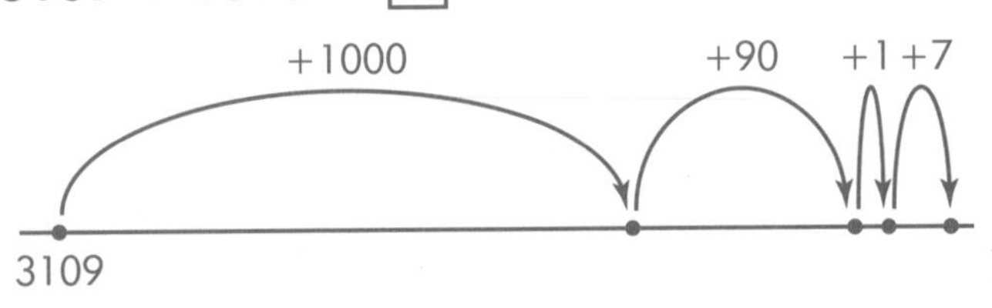

1124 + 7122 = ☐

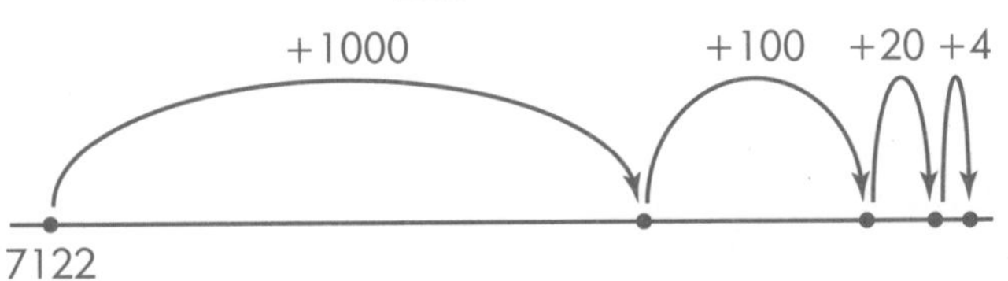

94 – 77 = ☐

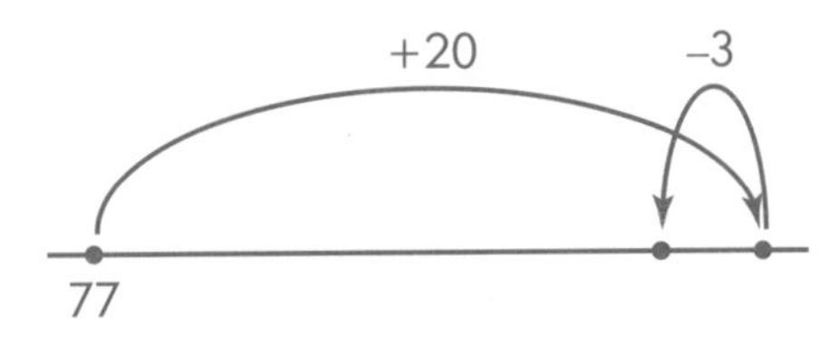

184 – 92 = ☐

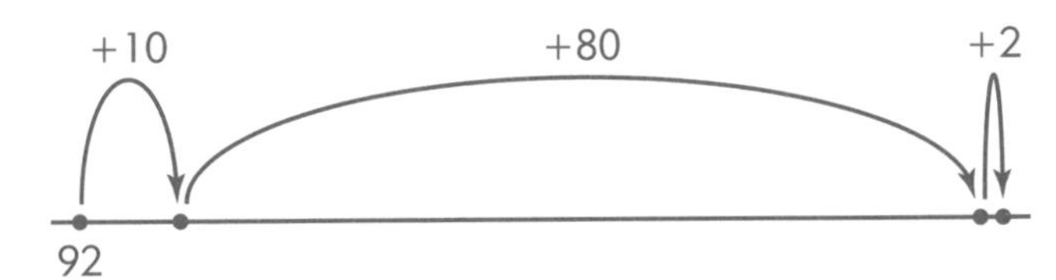

708 – 518 = ☐

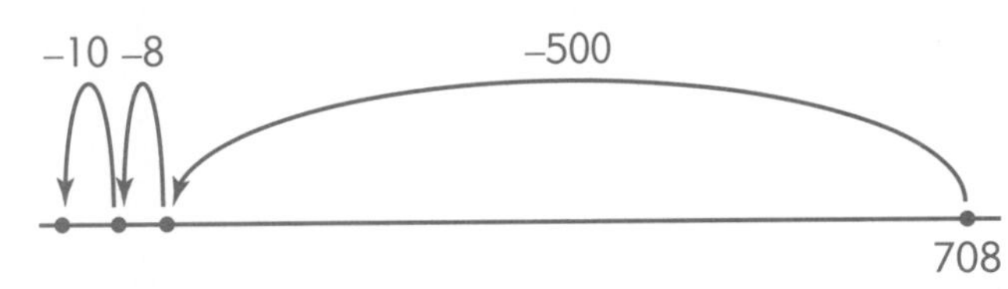

746 – 349 = ☐

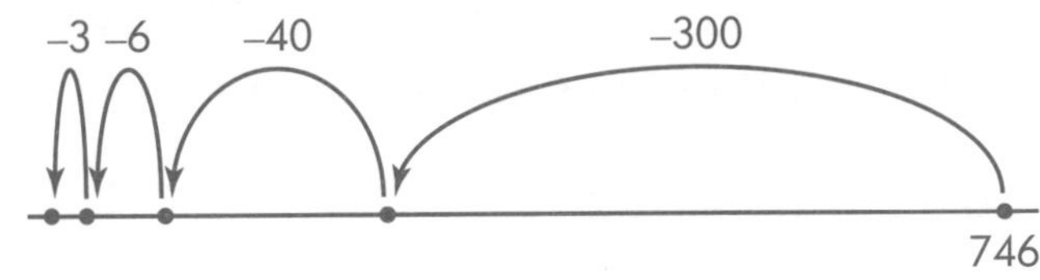

1034 – 587 = ☐

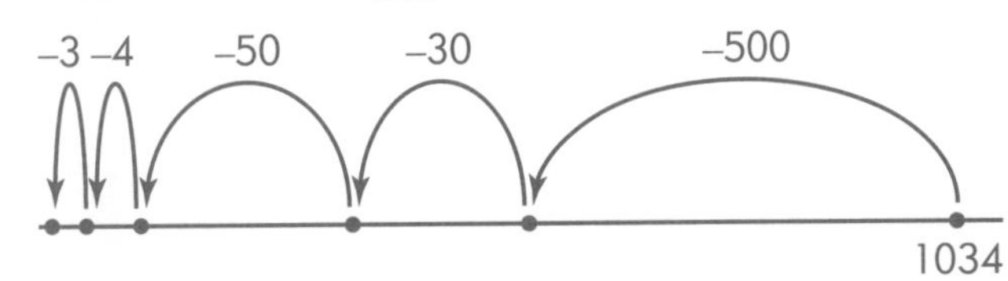

9000 – 7848 = ☐

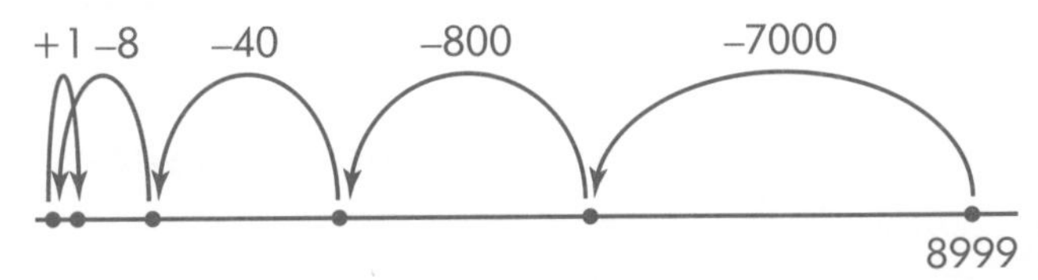

see Student Book page 43

Position on a grid

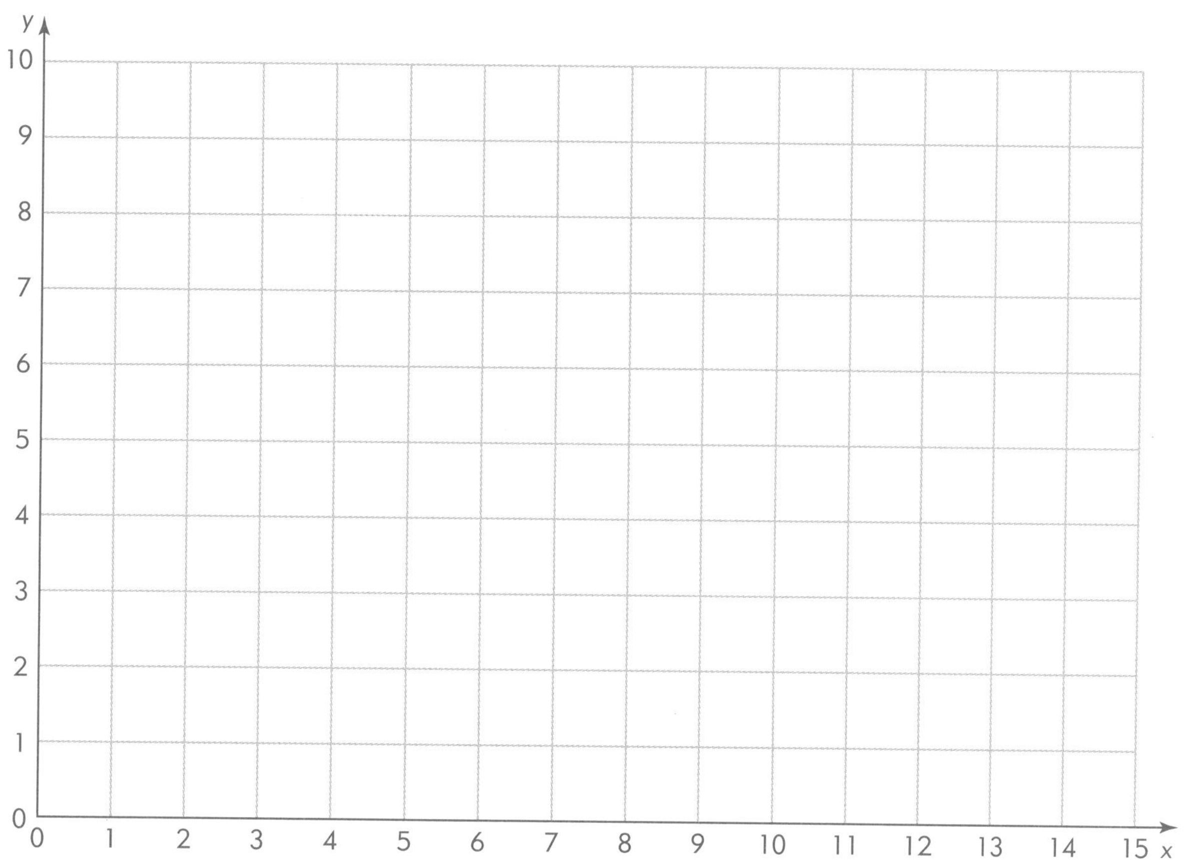

Plot and label each set of points.

Draw lines to join them up (in order).

Write the name of each shape.

1 A(1, 10), B(4, 7), C(1, 7) ______________

2 D(12, 10), E(14, 10), F(14, 3) G(12, 3) ______________

3 H(8, 10), I(11, 7), J(8, 4), K(5, 7) ______________

4 L(2, 1), M(6, 1), N(6, 3) O(4, 5), P(2, 3) ______________

5 Q(9, 3), R(9, 0), S(13, 0) ______________

see Student Book page 47

Multiplication facts

Use these tests to check how well you know your multiplication facts.

Write the answers only.

Time how long it takes you to complete each test.

Test 1	Test 2	Test 3
$4 \times 5 =$	$2 \times 10 =$	$2 \times 10 =$
$3 \times 8 =$	$1 \times 8 =$	$9 \times 9 =$
$9 \times 2 =$	$4 \times 4 =$	$9 \times 8 =$
$1 \times 9 =$	$3 \times 6 =$	$10 \times 9 =$
$4 \times 6 =$	$7 \times 10 =$	$4 \times 3 =$
$3 \times 7 =$	$3 \times 9 =$	$5 \times 9 =$
$2 \times 7 =$	$8 \times 9 =$	$8 \times 6 =$
$5 \times 5 =$	$5 \times 3 =$	$5 \times 8 =$
$2 \times 8 =$	$9 \times 4 =$	$9 \times 3 =$
$9 \times 3 =$	$7 \times 5 =$	$6 \times 7 =$
$6 \times 5 =$	$10 \times 9 =$	$5 \times 4 =$
$10 \times 7 =$	$8 \times 8 =$	$1 \times 9 =$
$7 \times 6 =$	$7 \times 7 =$	$8 \times 4 =$
$3 \times 9 =$	$9 \times 9 =$	$6 \times 9 =$
$5 \times 8 =$	$6 \times 8 =$	$6 \times 7 =$
$4 \times 9 =$	$10 \times 6 =$	$7 \times 8 =$
$6 \times 6 =$	$7 \times 9 =$	$4 \times 9 =$
$5 \times 10 =$	$9 \times 6 =$	$6 \times 5 =$
$6 \times 9 =$	$8 \times 5 =$	$6 \times 10 =$
$2 \times 3 =$	$9 \times 7 =$	$4 \times 9 =$
Time:	Time:	Time:

see Student Book pages 49 and 50

Division facts

Use these tests to check how well you know your division facts.

Write the answers only.

Time how long it takes you to complete each test.

Test 1	Test 2	Test 3
16 ÷ 2 =	21 ÷ 3 =	12 ÷ 4 =
30 ÷ 3 =	36 ÷ 4 =	32 ÷ 8 =
24 ÷ 8 =	90 ÷ 9 =	27 ÷ 9 =
27 ÷3 =	24 ÷ 4 =	80 ÷ 8 =
45 ÷ 5 =	81 ÷ 9 =	64 ÷ 8 =
81 ÷ 9 =	25 ÷ 5 =	48 ÷ 8 =
25 ÷ 5 =	15 ÷ 3 =	45 ÷ 5 =
24 ÷ 4 =	64 ÷ 8 =	56 ÷ 8 =
48 ÷ 6 =	18 ÷ 6 =	12 ÷ 2 =
50 ÷ 5 =	42 ÷ 6 =	28 ÷2 =
72 ÷ 9 =	63 ÷ 7 =	72 ÷ 8 =
48 ÷ 8 =	20 ÷ 4 =	35 ÷ 5 =
30 ÷ 6 =	49 ÷ 7 =	8 ÷ 1 =
35 ÷ 7 =	72 ÷ 9 =	36 ÷ 6 =
64 ÷ 8 =	50 ÷ 5 =	14 ÷ 2 =
63 ÷9 =	54 ÷ 9 =	18 ÷ 9 =
54 ÷ 6 =	16 ÷ 2 =	48 ÷ 6 =
24 ÷ 8 =	28 ÷ 7 =	36 ÷ 9 =
90 ÷ 10 =	36 ÷ 6 =	48 ÷ 6 =
21 ÷ 3 =	35 ÷ 7 =	54 ÷ 9 =
Time:	Time:	Time:

see Student Book pages 49 and 50

Factors

1 Find the factor pairs for each number. Write them in the towers.

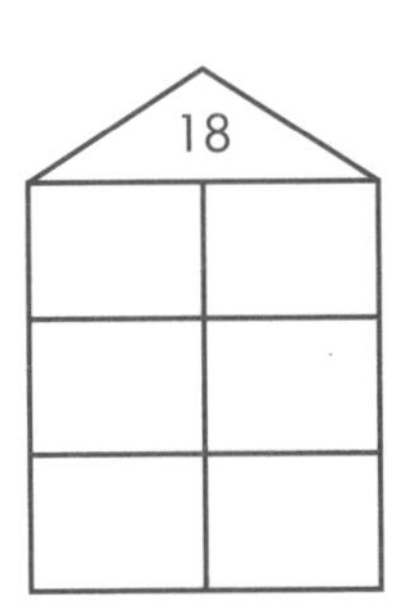

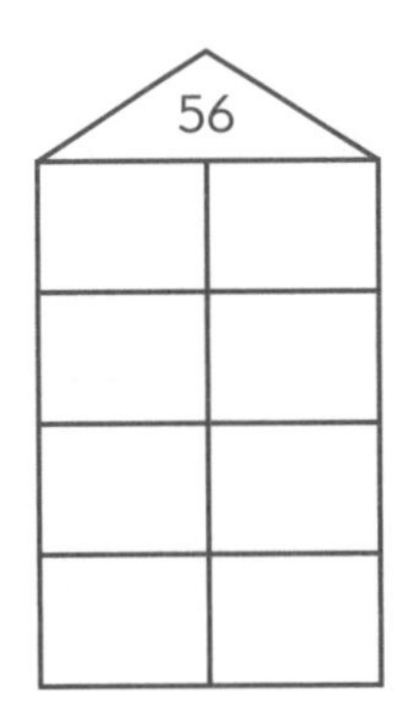

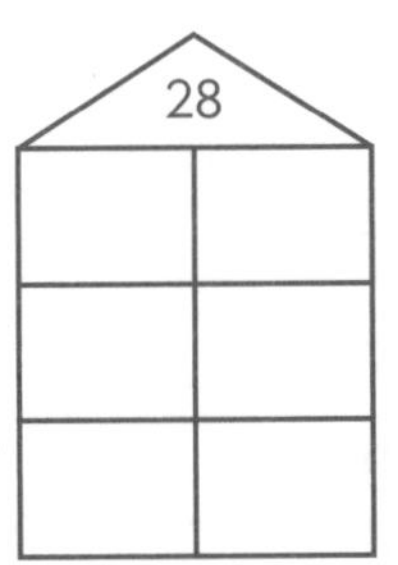

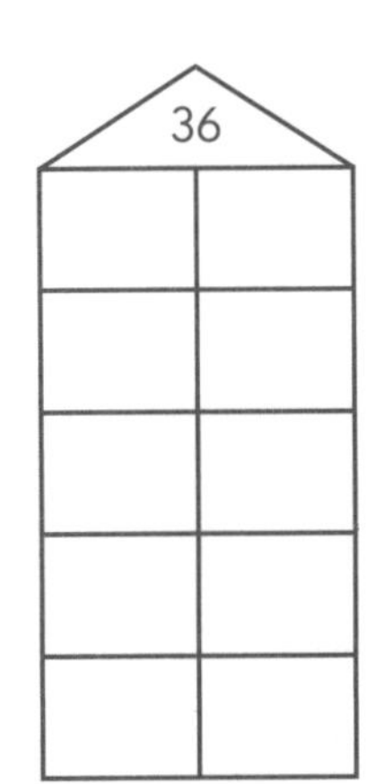

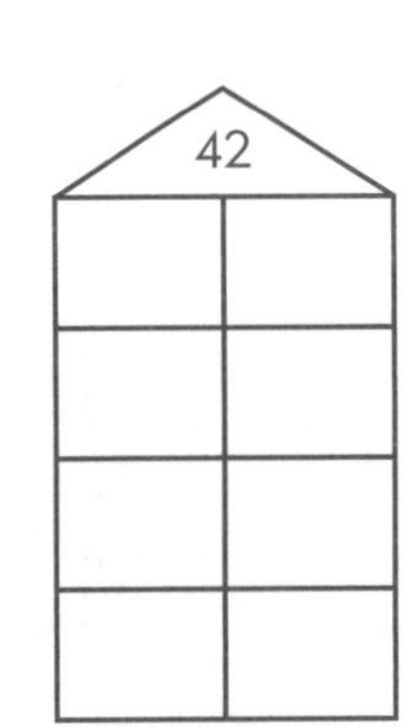

60

2 Complete these tables.

Product	12	18	36	48	30	100	90	81
Factor	1	9	6	6	5	10	9	9
Factor								

Factor	1	3	4	5	6	7	8	8
Factor								
Product	8	27	32	45	42	49	64	72

see Student Book page 54

Divisibility rules

1 Write the numbers in the box into the correct columns in the table.

17 28 19 16 113 29 45 58 109 124

148 1120 900 2356 650 191 1190 4512 4509 5201

Divisible by 2	Not divisible by 2

2 Circle all the numbers that are divisible by 5.

Colour in the circles if the number is also divisible by 10.

43 45 50 66 60 75 80 85 100

125 140 142 188 185 190 200 305 310

15 19 150 180 455 500 505 515 672

625 675 800 923 940 965 1000 1025 1099

3 Circle all the numbers that are divisible by 100.

30 300 350 3000 3500 3080 3800

120 1200 1250 1020 12 000 12 500 12 050

400 800 480 840 8400 4800 48 800 84 400

4 Are all numbers that are divisible by 100 also divisible by:

a 2 **b** 5 **c** 10

Explain how you decided.

see Student Book page 57

Different scales

This table gives the prices per kilogram of bananas at five different stores

Store	Farmers-R-us	Fancy Fruits	Cheap-n-Fresh	Fresh Things	Speciality Fruits
Price per kilogram ($)	2.50	4.00	1.00	2.00	6.00

1 Use the data in the table to complete these bar graphs comparing the prices of the different items.

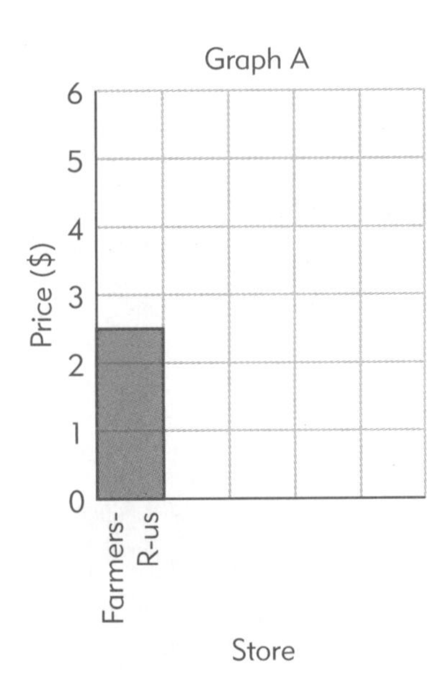

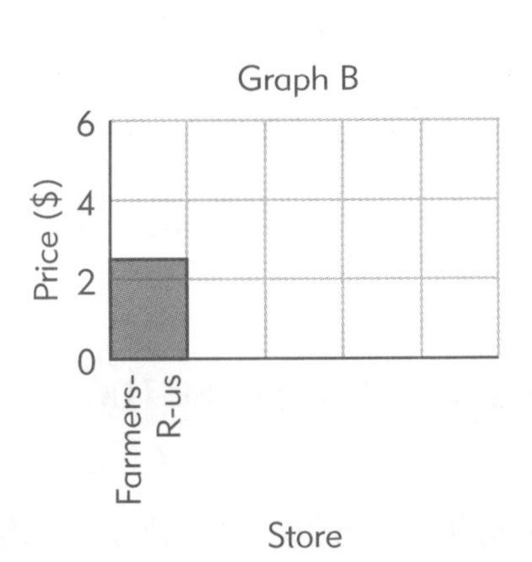

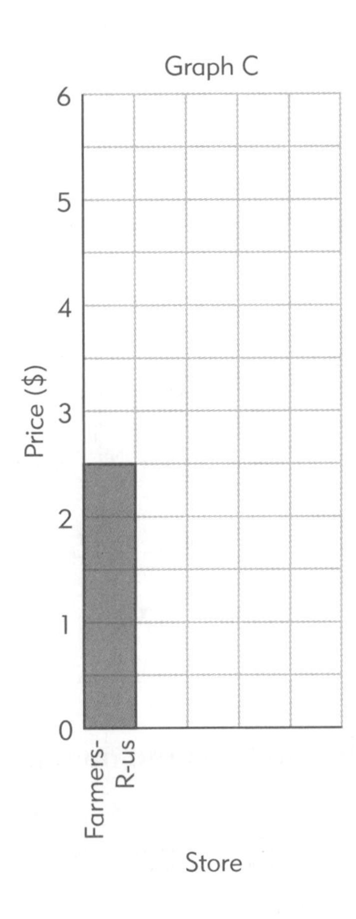

2 Discuss these questions, with a partner.

a Do all three graphs show the same information?

b What makes the graphs look different?

c Which graph do you think gives the most accurate picture of the data? Why?

see Student Book page 59

Show choices on a pictogram

Use this grid to complete question 1.

Abseiling	
Canoeing	
Wind-surfing	
Pony-trekking	
Parachuting	
Archery	
Orienteering	
Climbing	
Dinghy sailing	
Water-skiing	

Key
=

Use this table to record your data.

Activity	Tally	Activity	Tally
Abseiling		Archery	
Canoeing		Orienteering	
Wind-surfing		Climbing	
Pony-trekking		Dinghy sailing	
Parachuting		Water-skiing	

Use this grid to draw the pictogram for your class.

Abseiling	
Canoeing	
Wind-surfing	
Pony-trekking	
Parachuting	
Archery	
Orienteering	
Climbing	
Dinghy sailing	
Water-skiing	

Key
=

see Student Book pages 60 and 61

Positive and negative numbers

1 Fill in the missing numbers on each number line.

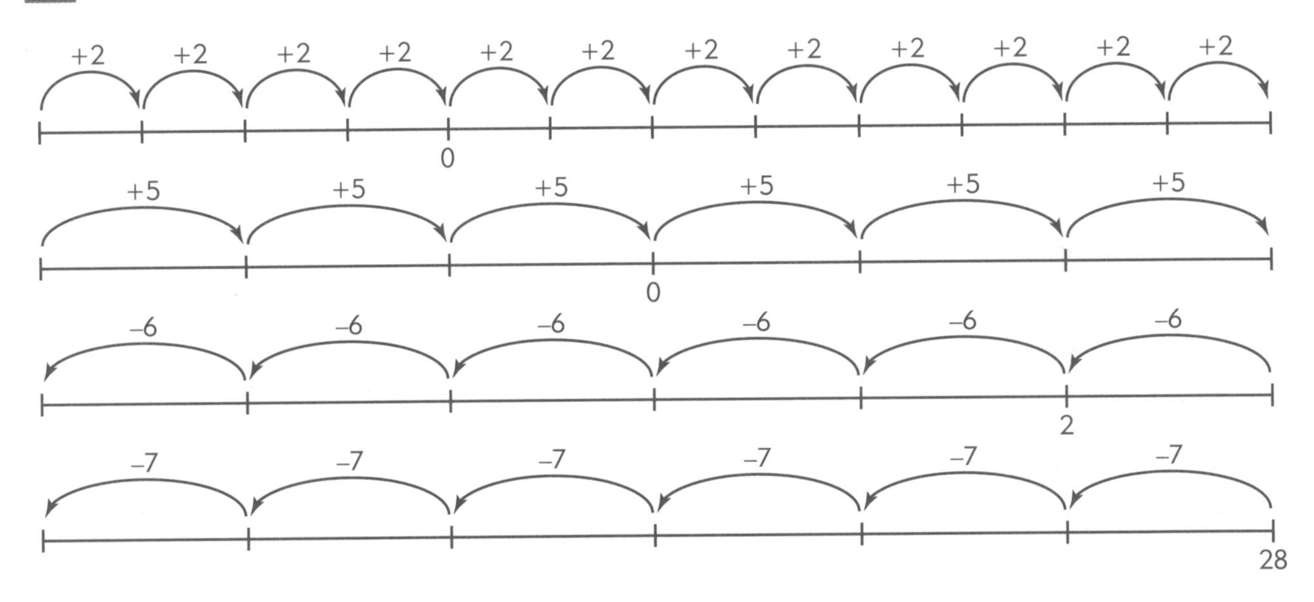

2 Write the number shown by each arrow on the number lines.

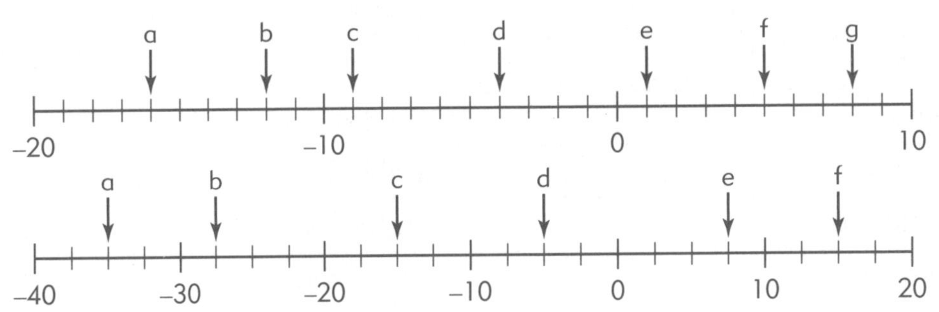

3 Estimate and show where you think each number would go on the given number lines.

a 0, –2, 1, 4, –3

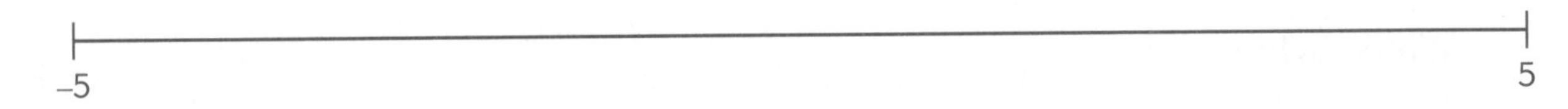

b –1, –3, –5, –9, 1

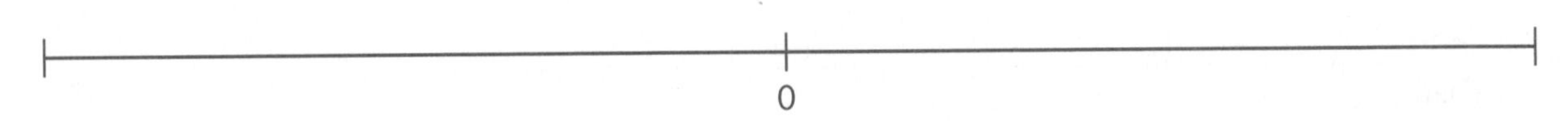

c –15, 10, –5, 0, –10

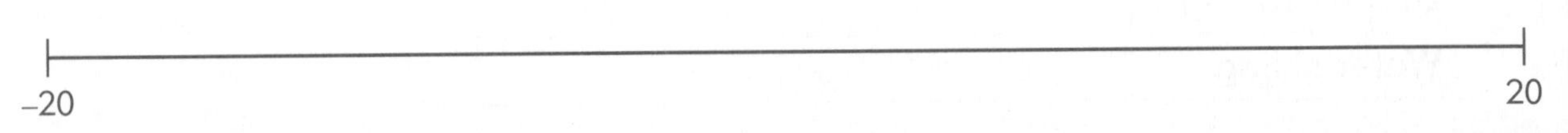

see Student Book page 64

Temperature changes

Colour the thermometers to show the given temperatures.

Write what the temperature would be if it changed as given.

15°C

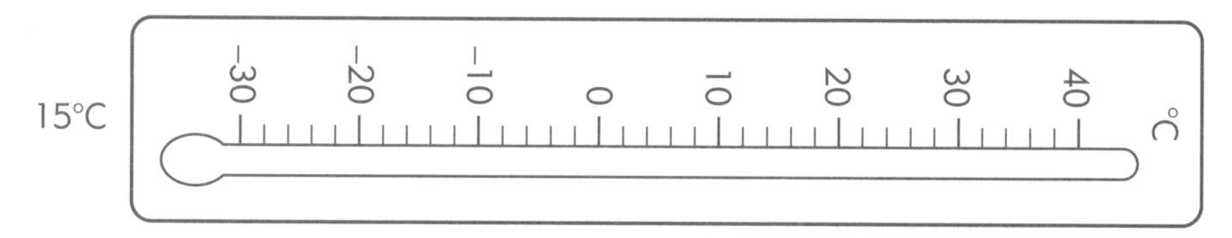

4 degrees warmer

12 degrees colder

12°C

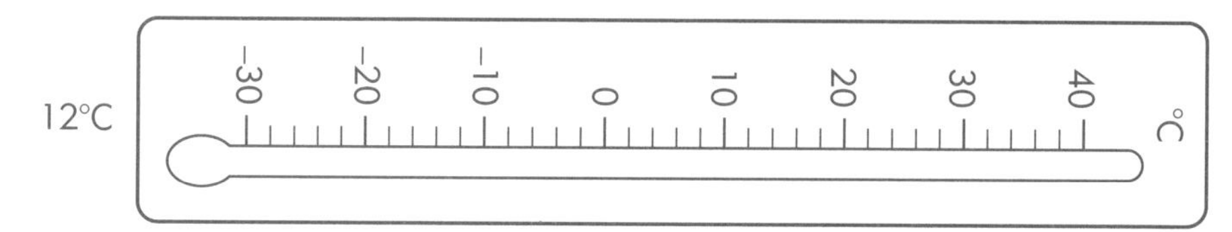

9 degrees colder

Another 3 degrees colder

19°C

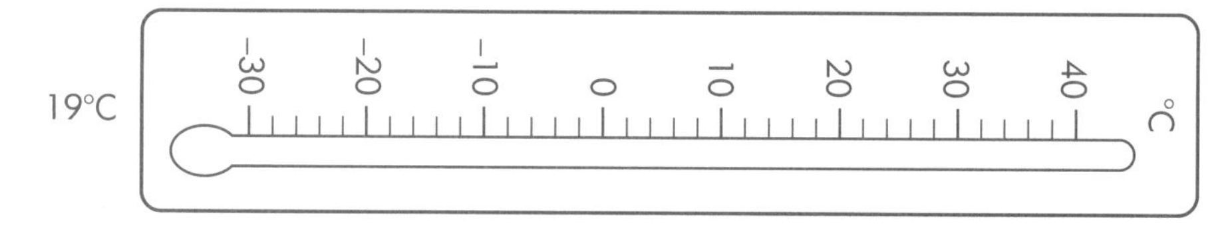

5 degrees warmer

22 degrees colder

21°C

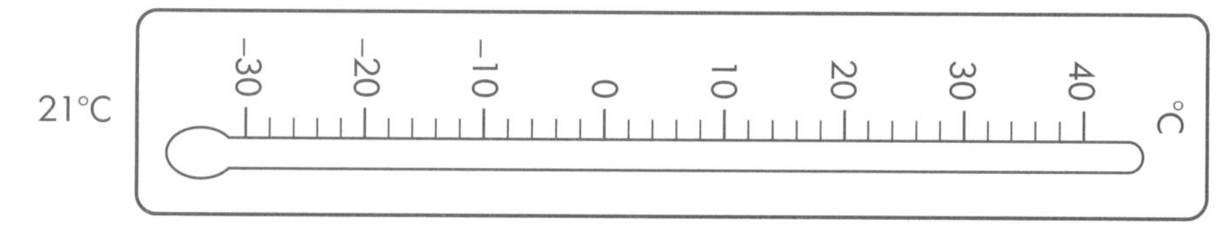

A rise of 15 °C

A decrease of 18 °C

–4°C

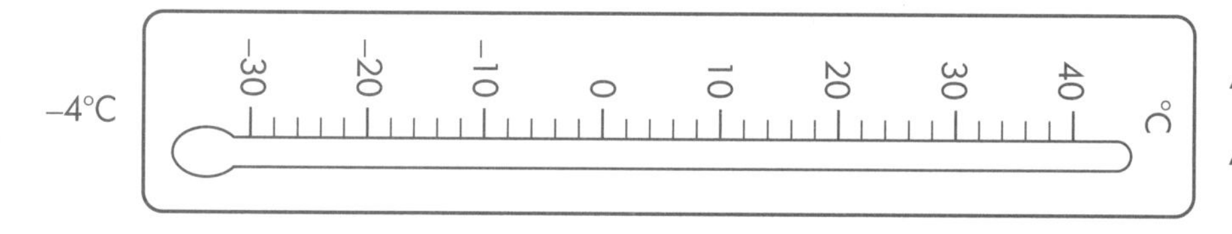

A drop of – 6 °C

A rise of 8 °C

–2°C

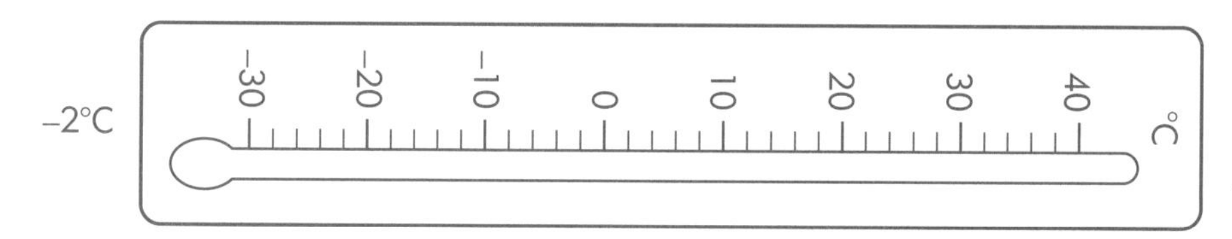

9 °C colder

Another 3 °C colder

see Student Book page 66

Looking at parallel lines

Look at these pictures. Tick the ones that have parallel lines.

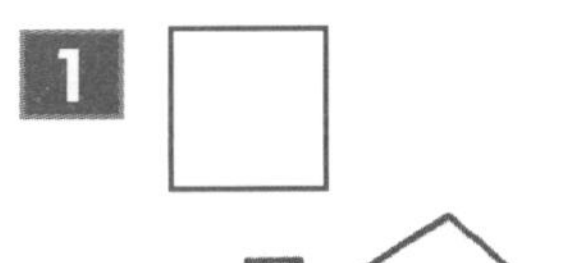

1

2

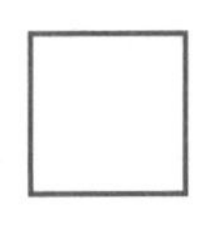

3

4

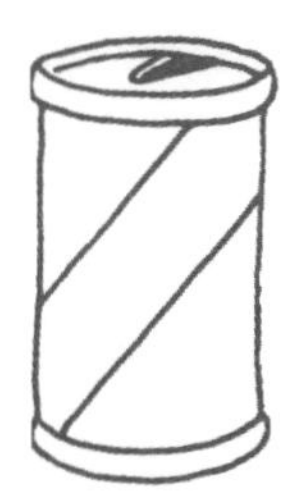

5

6

7

8

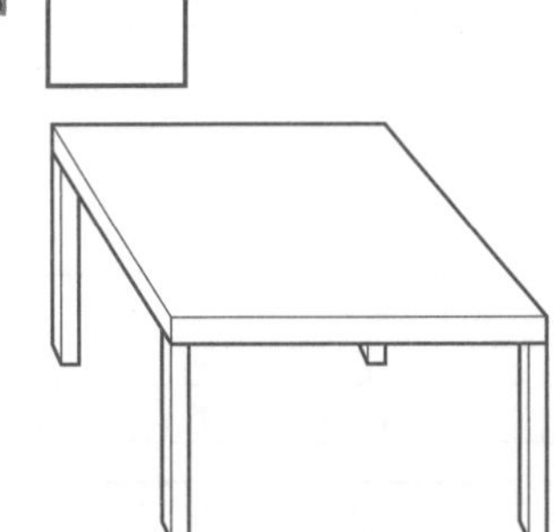

9

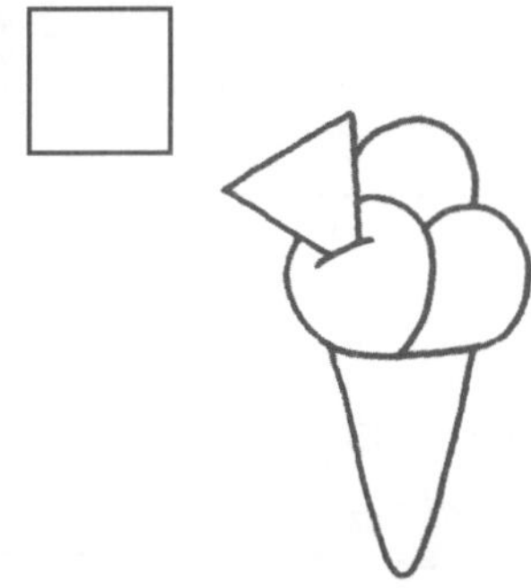

10

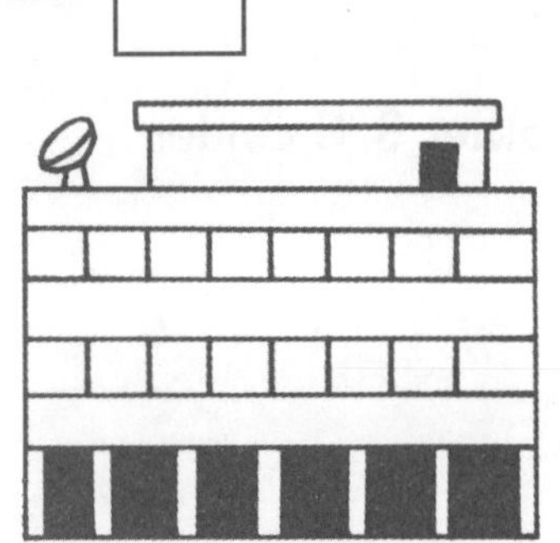

11

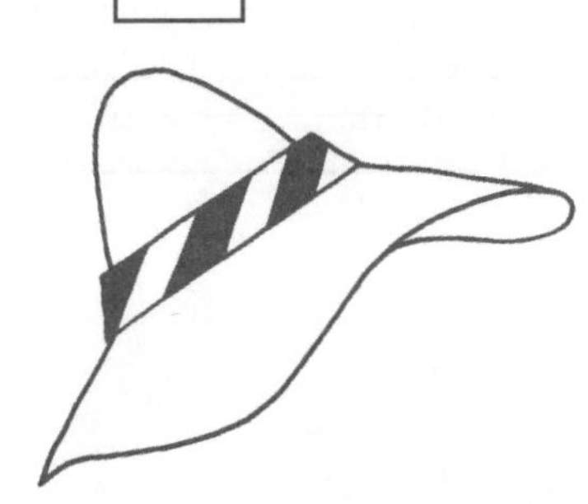

12

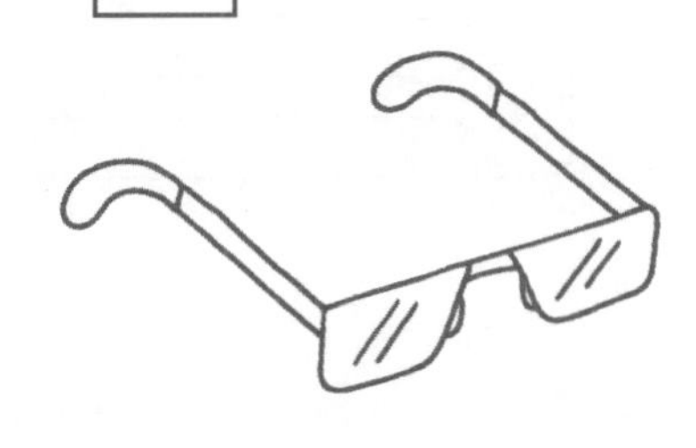

see Student Book page 70

Parallel and perpendicular lines in real life

1 **On this five-barred gate, colour in:**

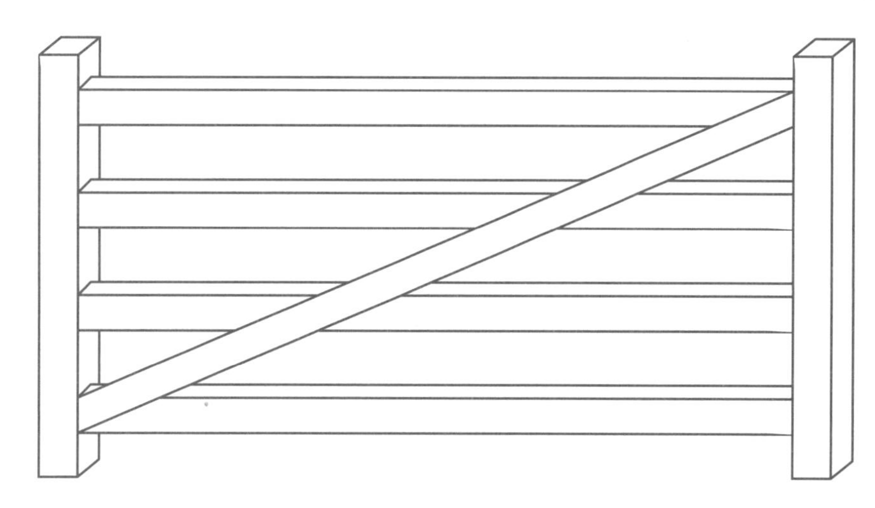

a one pair of horizontal parallel bars green

b one pair of vertical parallel bars red

c the diagonal blue.

2 **Write the size of the angles formed between the diagonal bar and the other bars.**

____

3 **What do you notice about the angles formed by a line intersecting parallel lines?**

____

____

4 **Draw your own sets of lines:**

a one pair of diagonal parallel lines

b one pair of perpendicular lines.

5 **What can you say about the two angles formed by a pair of perpendicular lines?**

____

____

see Student Book page 71

Angles

1 **On the bicycle:**

a Find five acute angles. Mark them blue.

b Find three obtuse angles. Mark them green.

c Find four right angles. Mark them red.

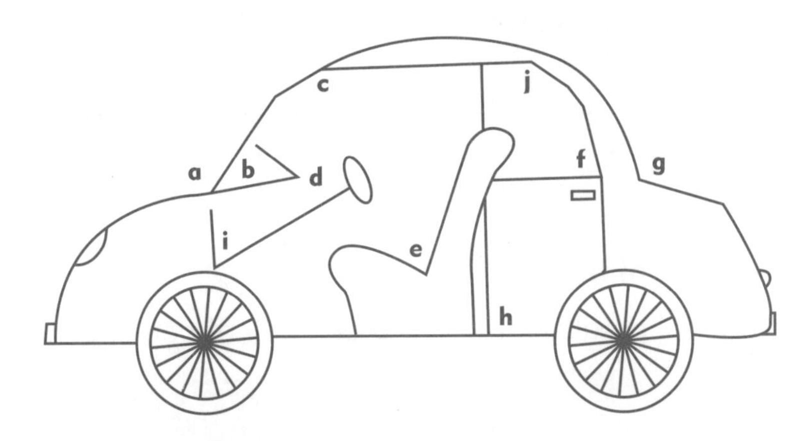

2 **Write whether each numbered angle in this drawing is acute, obtuse, or a right angle.**

a ______________________ **b** ______________________

c ______________________ **d** ______________________

e ______________________ **f** ______________________

g ______________________ **h** ______________________

i ______________________ **j** ______________________

see Student Book page 74

Angles on a straight line

The line AB is a straight line in each diagram.

Use the size of the given angles to calculate the size of the unmarked angle in each diagram.

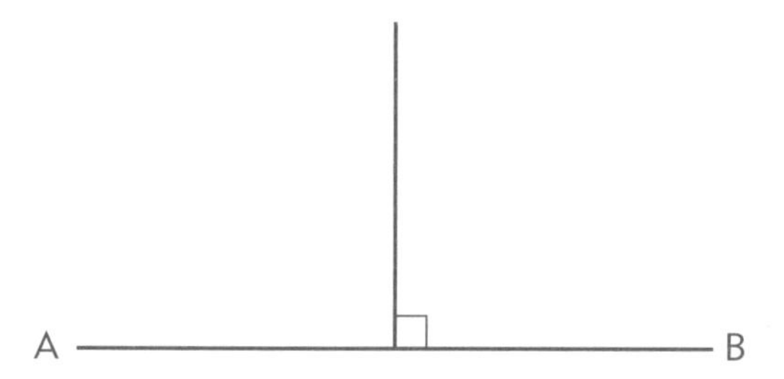

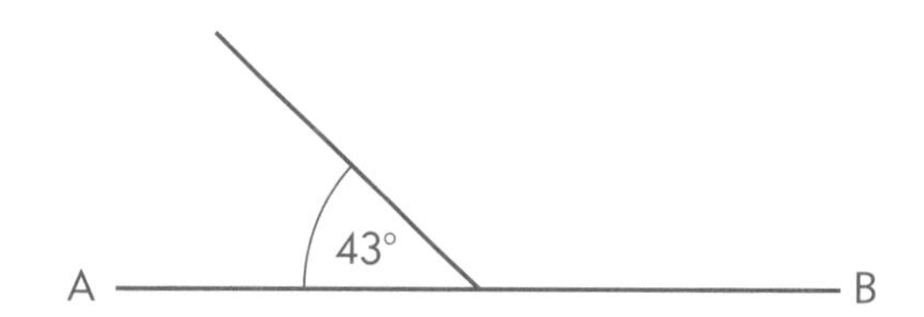

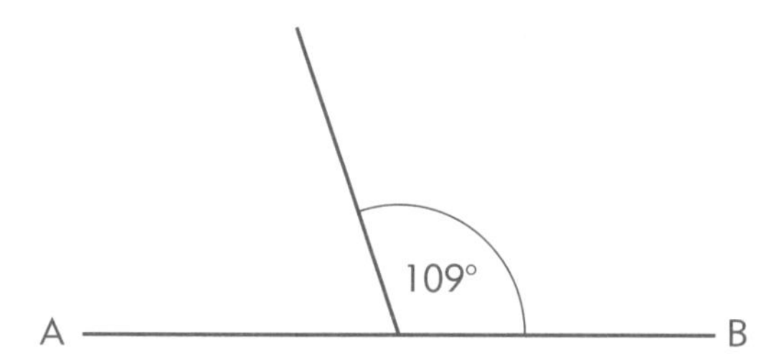

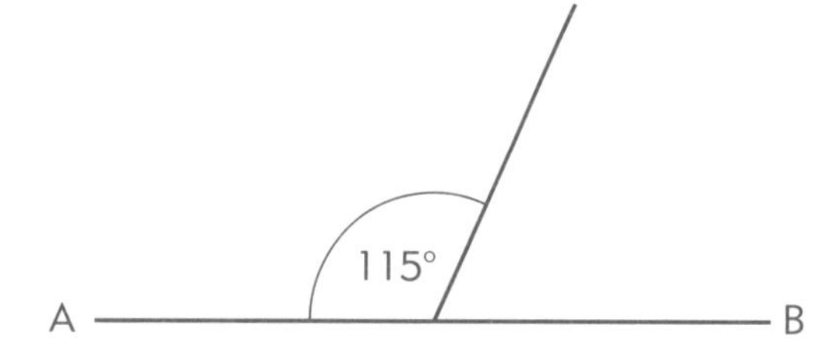

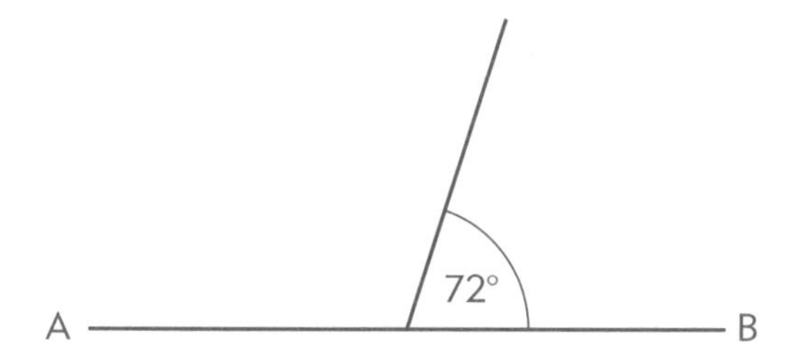

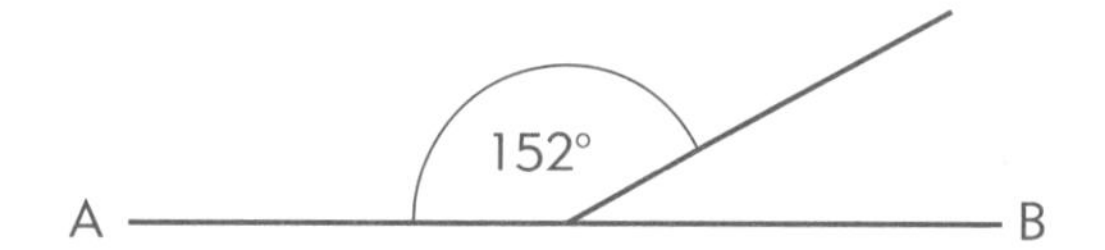

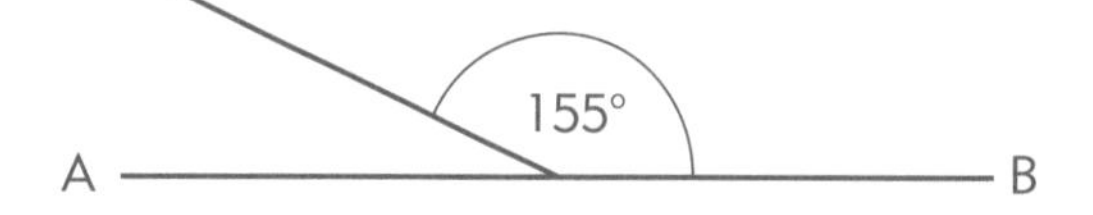

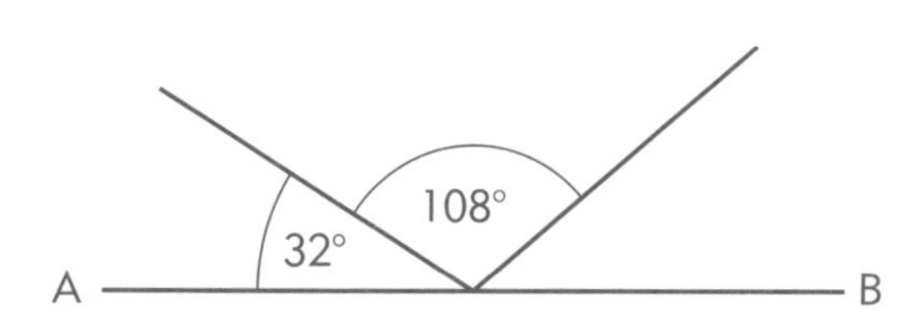

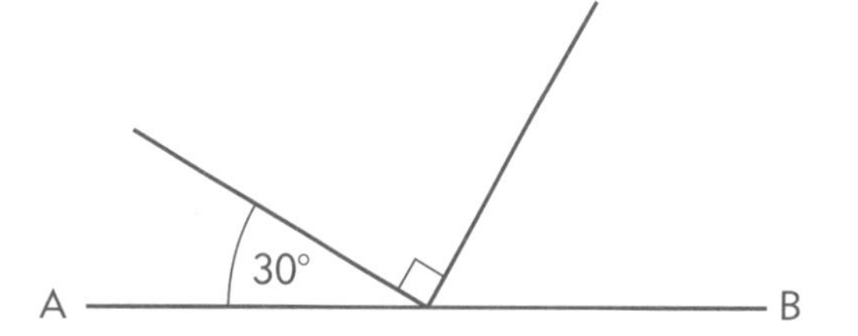

see Student Book page 75

Equivalent fractions

Use the circles to help you answer the questions.

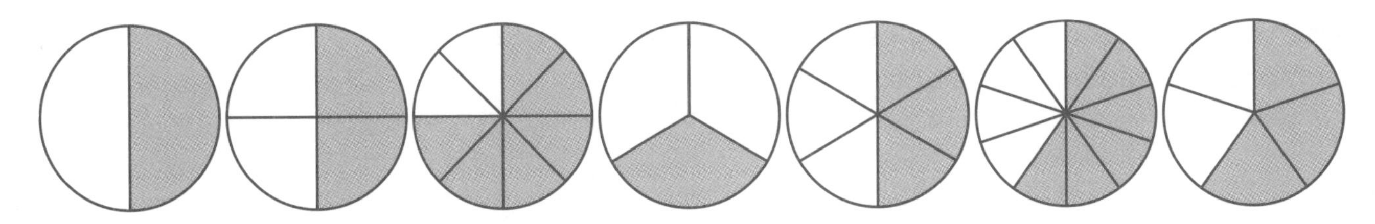

1 Write the equivalent fractions:

a $\frac{1}{4} = \frac{\square}{8}$　　**b** $\frac{2}{5} = \frac{\square}{10}$　　**c** $\frac{3}{4} = \frac{6}{\square}$

d $\frac{1}{2} = \frac{5}{\square}$　　**e** $\frac{6}{10} = \frac{\square}{5}$　　**f** $\frac{1}{3} = \frac{\square}{6}$

2 Circle the fraction that is not equivalent in each set.

a $\frac{1}{2}$　$\frac{4}{8}$　$\frac{2}{5}$　$\frac{5}{10}$

b $\frac{1}{3}$　$\frac{2}{6}$　$\frac{3}{9}$　$\frac{3}{10}$

c $\frac{3}{4}$　$\frac{6}{8}$　$\frac{8}{10}$　$\frac{9}{12}$

3 Circle the fractions that are less than $\frac{1}{2}$.

$\frac{3}{10}$	$\frac{2}{8}$	$\frac{5}{6}$	$\frac{1}{6}$	$\frac{3}{6}$	$\frac{3}{8}$	$\frac{9}{10}$	$\frac{4}{10}$
$\frac{1}{4}$	$\frac{2}{5}$	$\frac{5}{10}$	$\frac{4}{5}$	$\frac{6}{8}$	$\frac{4}{6}$	$\frac{8}{10}$	$\frac{4}{8}$

4 Write each set of fractions in order from smallest to greatest.

a $\frac{3}{4}$　$\frac{1}{4}$　$\frac{1}{2}$

b $\frac{1}{2}$　$\frac{9}{10}$　$\frac{4}{5}$　$\frac{3}{10}$

c $\frac{1}{2}$　$\frac{5}{8}$　$\frac{3}{8}$　$\frac{3}{4}$

d $\frac{2}{5}$　$\frac{1}{10}$　$\frac{1}{2}$　$\frac{3}{5}$　$\frac{4}{10}$

e $\frac{1}{3}$　$\frac{1}{2}$　$\frac{1}{5}$　$\frac{1}{10}$　$\frac{1}{8}$

5 Tick the statements that are true. Correct any statements that are false.

a $\frac{3}{5} = \frac{6}{10}$　　**b** $\frac{1}{2} > \frac{3}{4}$　　**c** $\frac{2}{8} < \frac{1}{4}$

d $\frac{5}{10} > \frac{3}{8}$　　**e** $\frac{5}{8} < \frac{1}{2}$　　**f** $\frac{3}{4} = \frac{6}{8}$

g $\frac{7}{10} < \frac{6}{8}$　　**h** $1 = \frac{10}{10}$　　**I** $\frac{5}{8} = \frac{3}{4}$

see Student Book page 77

More equivalent fractions

Remember: Equivalent fractions have the same value.

1 Complete this chart of equivalent fractions.

Thirds	Quarters	Fifths	Sixths	Eighths	Tenths
–	$\frac{2}{4}$	–	$\frac{3}{6}$	$\frac{4}{8}$	$\frac{5}{10}$
$\frac{1}{3}$					
$\frac{2}{3}$					
	$\frac{1}{4}$				
	$\frac{3}{4}$				
		$\frac{1}{5}$			
		$\frac{2}{5}$			
		$\frac{3}{5}$			
		$\frac{5}{5}$			

2 In a family of equivalent fractions, the simplest fraction is the one with the lowest numerator and denominator.

Use the chart to find the simplest equivalent fraction for each of these:

a $\frac{5}{10}$ ________

b $\frac{6}{8}$ ________

c $\frac{4}{6}$ ________

d $\frac{6}{10}$ ________

see Student Book page 77

Improper fractions and mixed numbers

Show each mixed number on the number line.

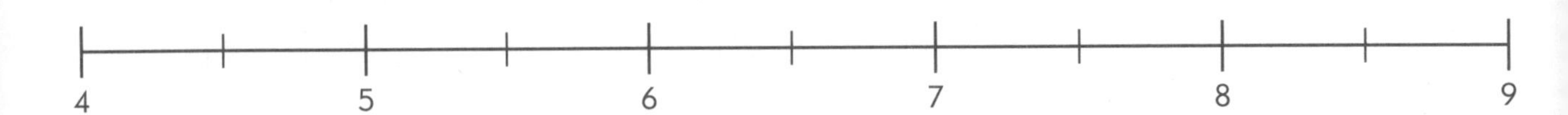

$8\frac{1}{2}$
$4\frac{1}{4}$
$6\frac{7}{10}$
$5\frac{1}{3}$
$8\frac{9}{10}$
$4\frac{3}{4}$
$5\frac{6}{8}$
$6\frac{1}{4}$
$7\frac{4}{5}$

see Student Book page 78

Name the fraction

1 **What fraction of the total set of shapes are:**

a triangles? ____________

b small triangles? ____________

c large triangles? ____________

d circles? ____________

e large circles? ____________

f small circles? ____________

g not circles? ____________

h squares? ____________

i large squares? ____________

j not squares? ____________

k hexagons? ____________

l shaded? ____________

m unshaded? ____________

n four-sided? ____________

o squares and hexagons? ____________

p unshaded hexagons? ____________

see *Student Book page 80*

Perimeter

Draw each of the following shapes.

1 A square with perimeter 16 cm.

2 A square with perimeter 20 cm.

3 A rectangle with perimeter 16 cm.

4 A rectangle with perimeter 20 cm.

see Student Book page 81

More perimeter and area

1 Here are some shapes on a 1 cm grid.

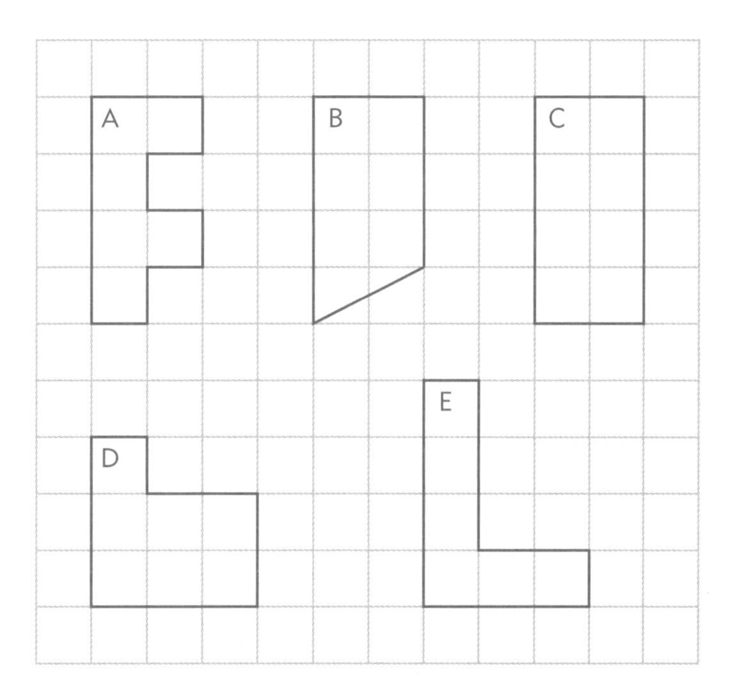

a Colour two shapes that have a perimeter of 12 cm.

b Which two shapes have a perimeter of 14 cm? _______ and _______

c What is the area of Shape B? _______

d Which has the greater area: Shape A or Shape E? _______

---

2 **The perimeter of this rectangle is 40 centimetres.**

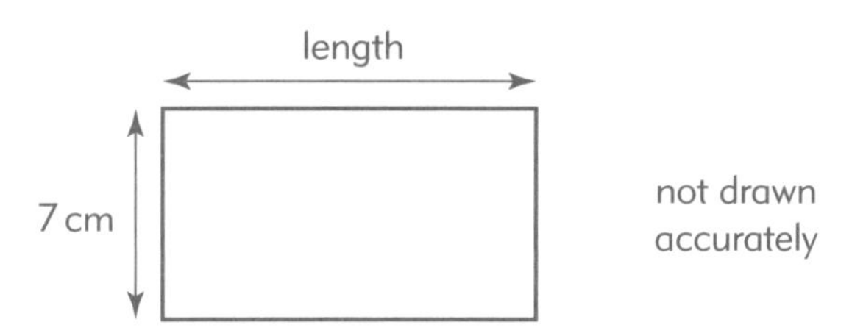

a Calculate the length of the rectangle. _______

b What is the area of this rectangle? _______

---

3 **What is the area of a square if its perimeter is:**

a 20 cm

b 64 cm

see Student Book page 84

Multiplying by 10 and 100

1 Fill in the missing numbers.

a 7 →(×10)→ ☐ →(×10)→ ☐ →(×10)→ ☐ →(×10)→ ☐

b 18 →(×10)→ ☐ →(×10)→ ☐ →(×100)→ ☐

c 97 →(×100)→ ☐ →(×10)→ ☐ →(×10)→ ☐

2 Work out the missing operations.

a 12 →☐→ 120 →☐→ 1200 →☐→ 120 000 →☐→ 1 200 000

b 23 →☐→ 2300 →☐→ 23 000 →☐→ 230 000

c 129 →☐→ 1290 →☐→ 129 000 →☐→ 1290 Think!

see Student Book page 85

Multiplying and dividing by 10 and 100

Complete this flow diagram.

Fill in one of the operations: × 10, × 100, ÷ 10 or ÷100 in each circle.

Write the correct numbers in the blocks.

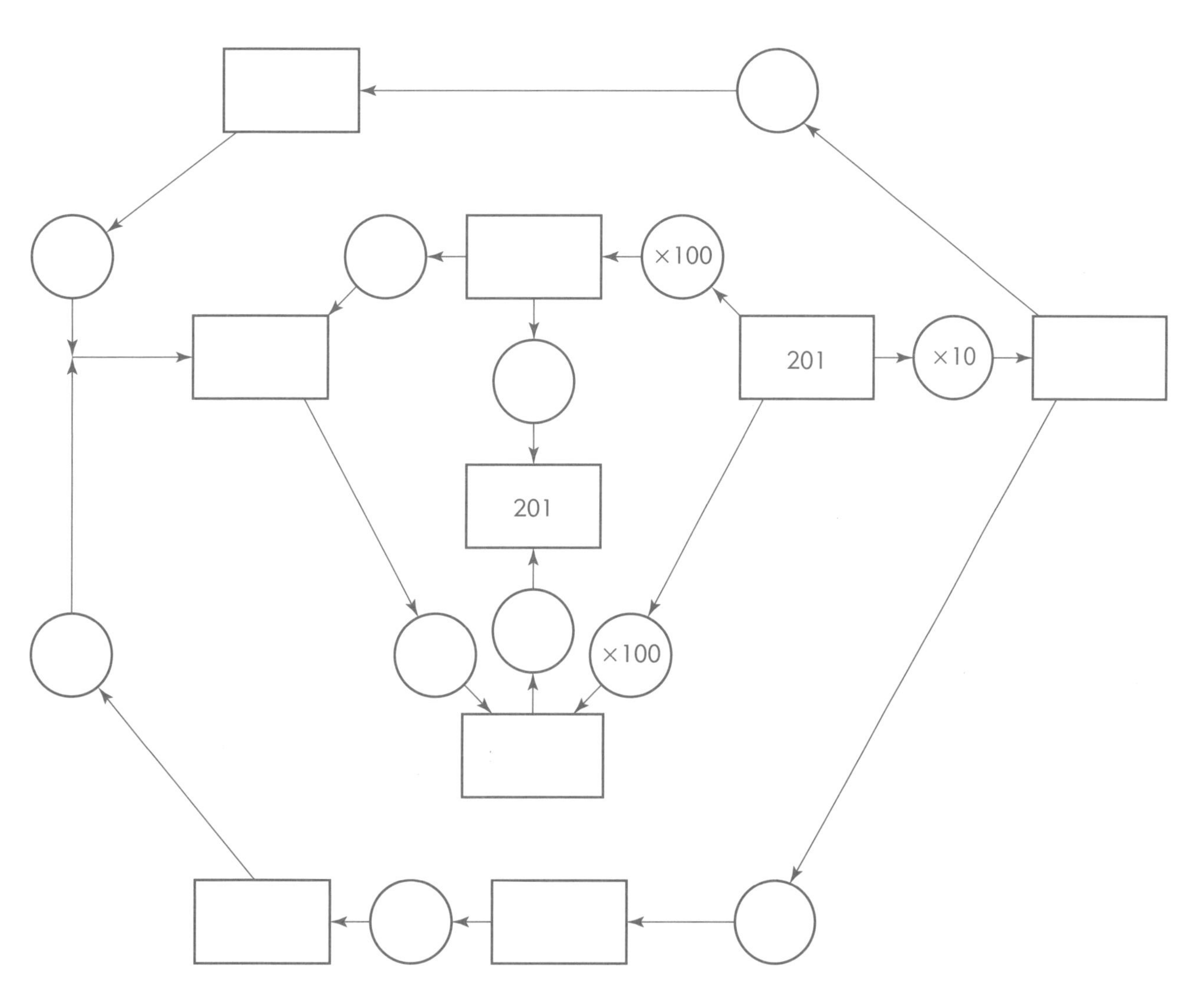

see Student Book page 86

Doubling and halving

1 Super electronics is having a half-price sale. Write the sale price of each item.

Was $128
Now:

Was $2700
Now:

Was $8300
Now:

---

2 The matching sides of the shapes on the right are double the length of those on the left. Write the missing lengths. Use a calculator to find the perimeter of all the shapes.

19
14 14
19
P =

P =

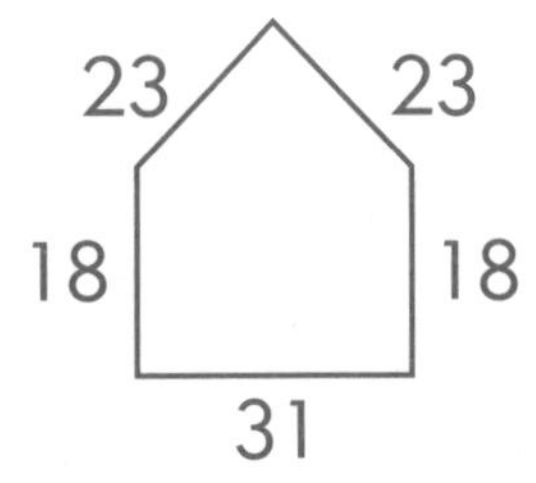

P =

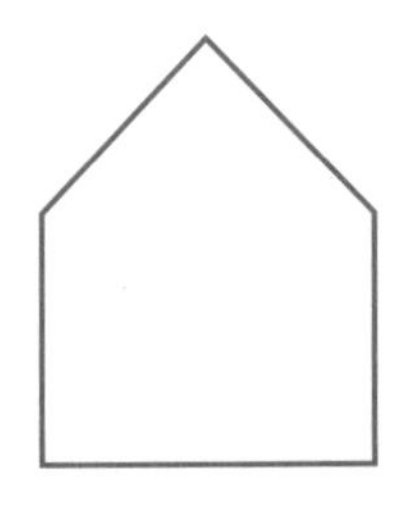

P =

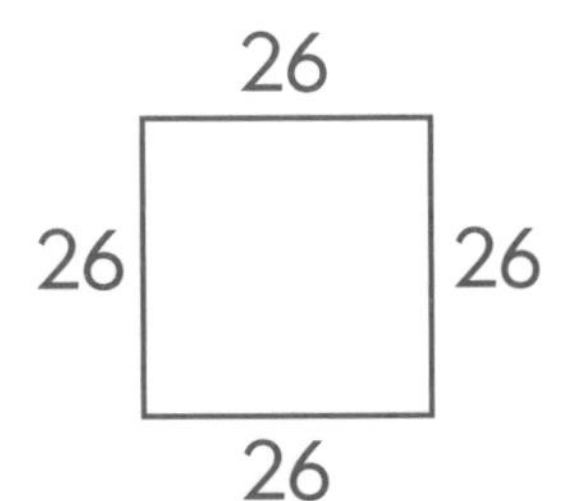

P =

P =

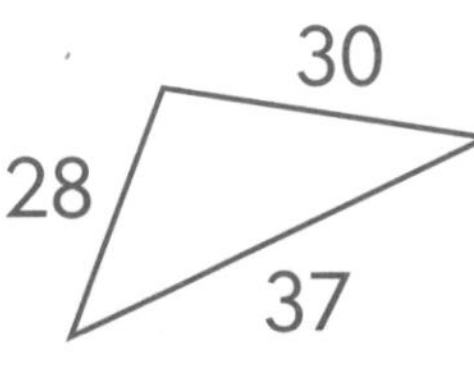

P =

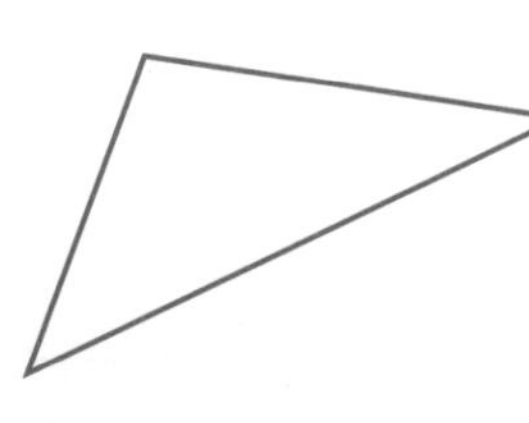

P =

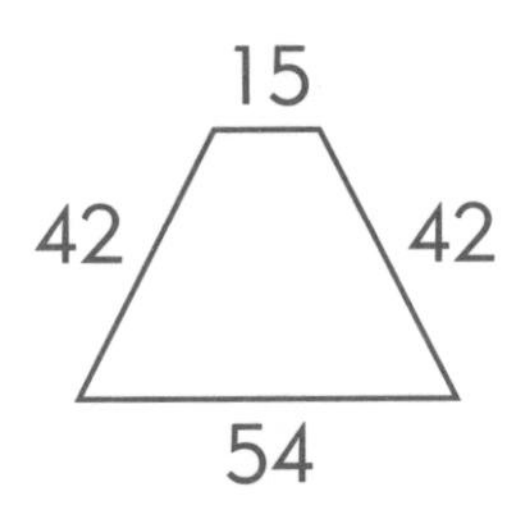

P =

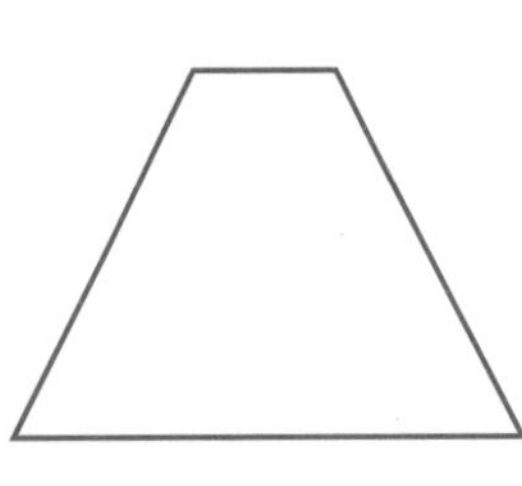

P =

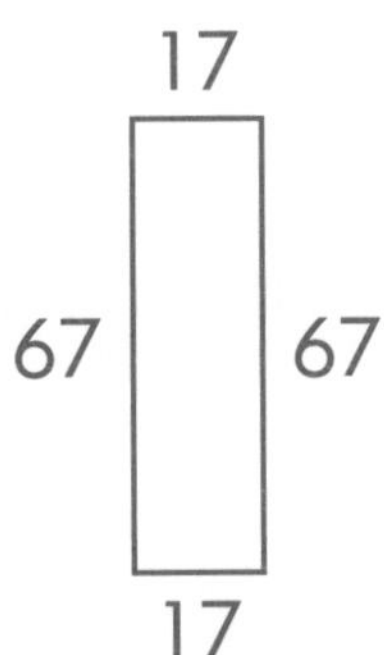

P =

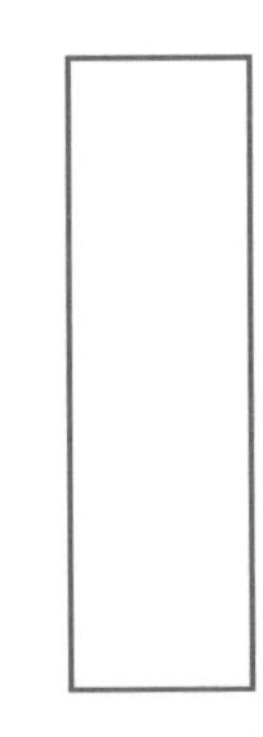

P =

see Student Book page 88

Reflections

Draw the reflection of each shape on the other side of the dotted line (or mirror line). Notice the equal amounts of space between the image and the mirror line, and the reflection and the mirror line.

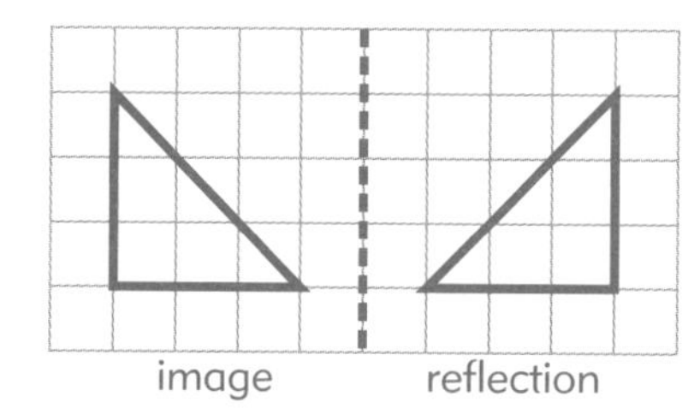

Remember: If there is space between the image and the mirror line, you need to leave the same amount of space between the reflection and the mirror line.

1

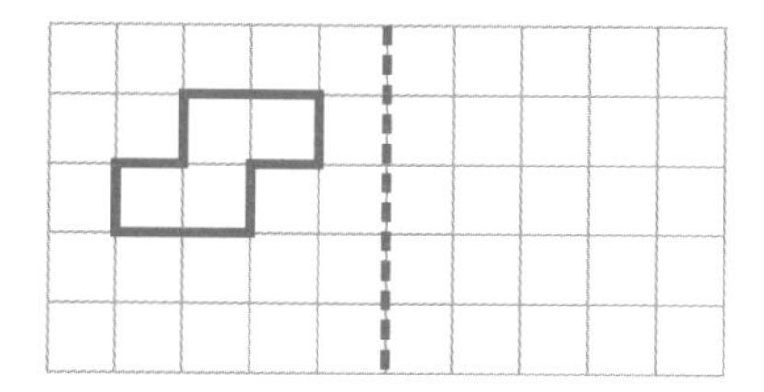

2

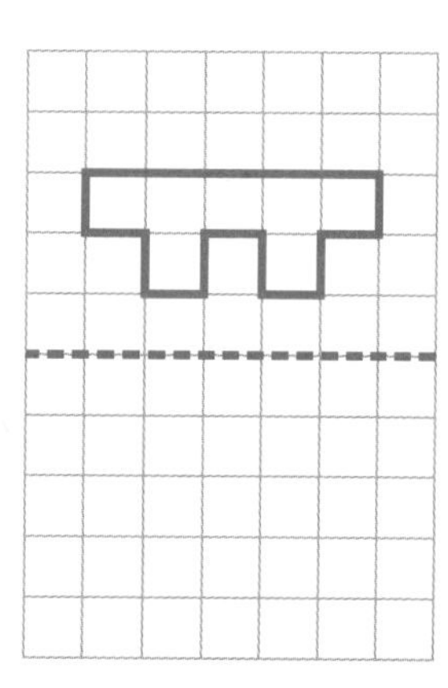

3

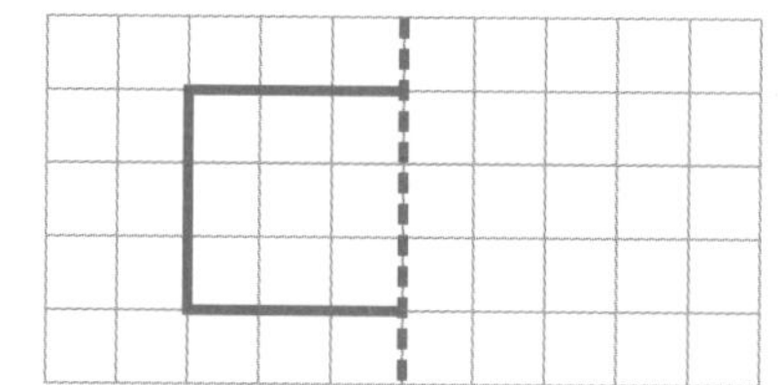

4

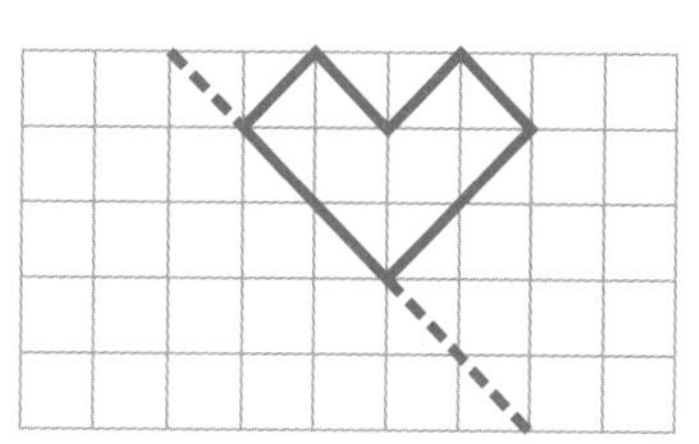

5

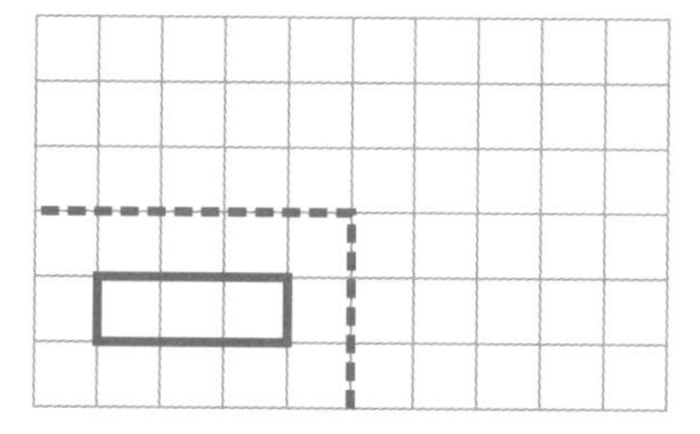

6

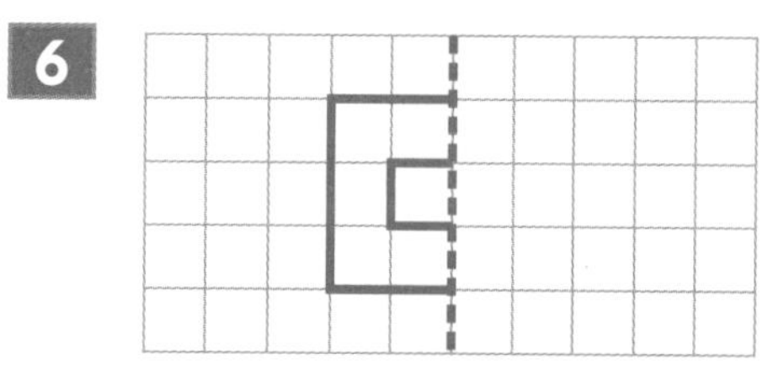

see Student Book page 92

Translated shapes

Draw each shape translated on the grid according to the instructions.

1 Translate the triangle forwards 4 blocks.

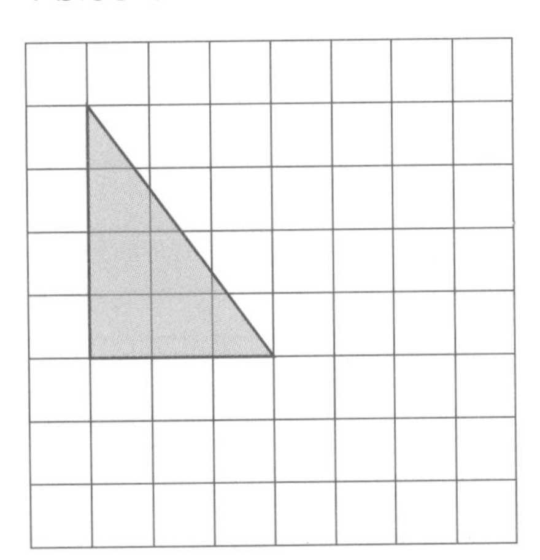

2 Slide the square down 3 blocks.

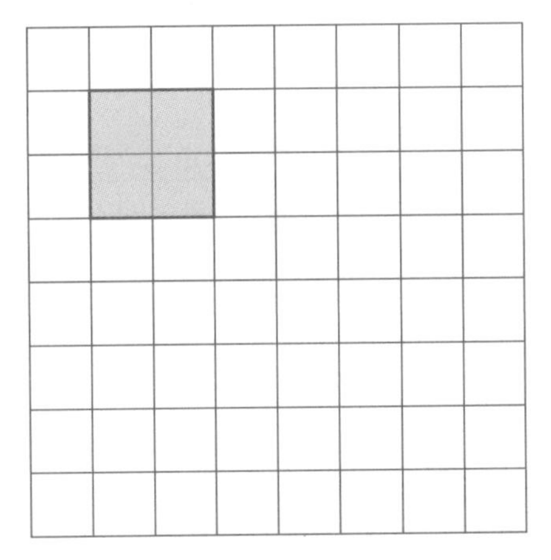

3 Translate this shape back 5 blocks.

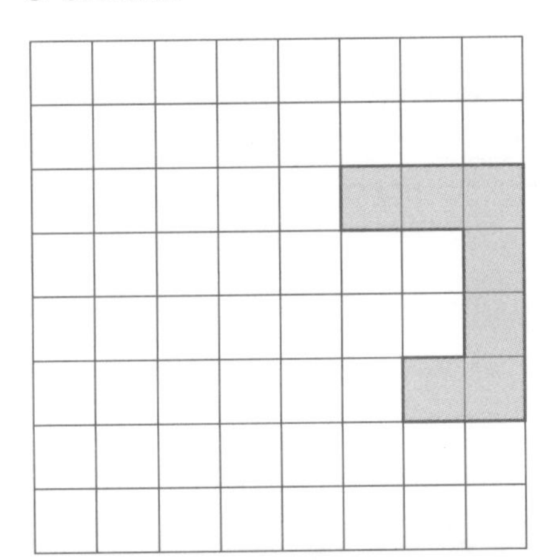

4 Slide this shape forwards 3 blocks.

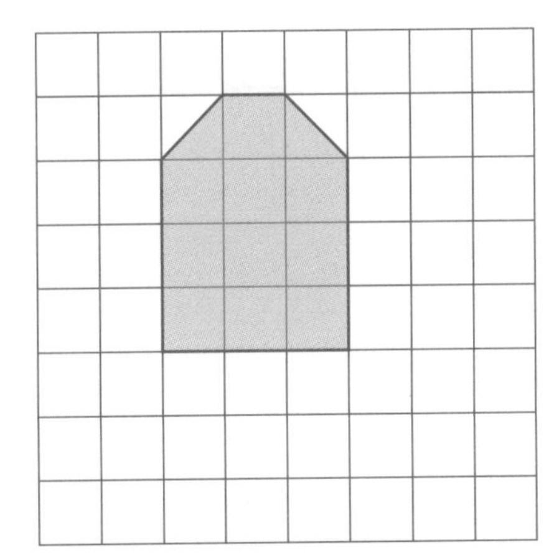

5 Translate this shape down 5 blocks and forwards 4 blocks.

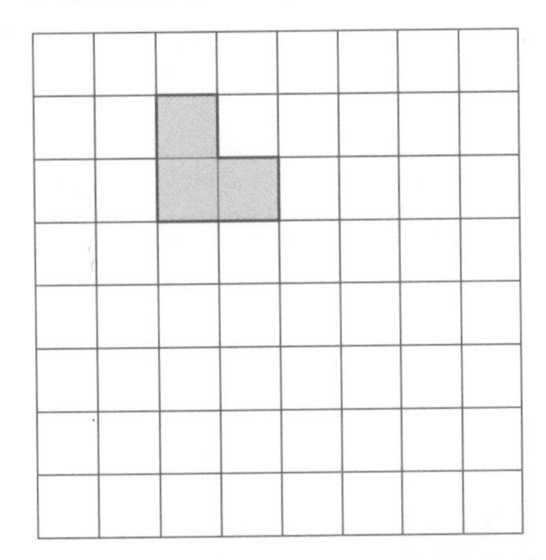

6 Slide this shape forwards 7 blocks and up 1 block.

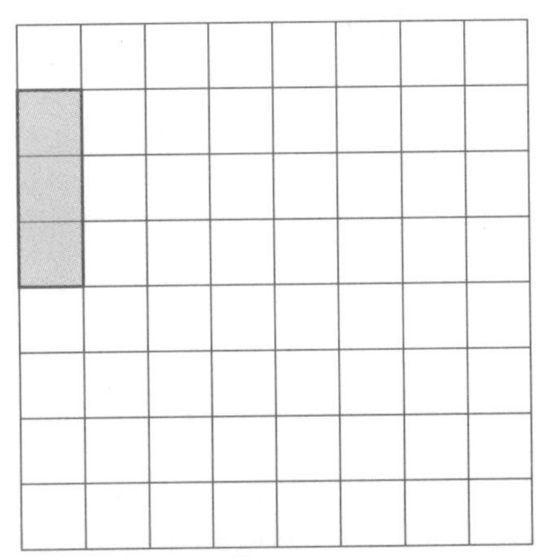

see Student Book page 94

Pairs of decimals that make 1

1 Find pairs of decimals that add up to 1.

As you find each pair, cross out the numbers.

0.8	0.45	0.9	0.65	0.19	0.76	0.05
0.81	0.77	0.15	0.75	0.68	0.5	0.87
0.62	0.49	0.38	0.35	0.32	0.13	0.97
0.25	0.3	0.4	0.58	0.98	0.14	0.51
0.99	0.71	0.59	0.17	0.95	0.79	0.23
0.41	0.2	0.29	0.84	0.24	0.1	0.21
0.4	0.86	0.5	0.41	0.6	0.42	0.65
0.5	0.85	0.03	0.7	0.16	0.01	0.59

2 List all the numbers that are not crossed out. Next to each one, write how much you would need to add to it to make 1.

see *Student Book page 96*

Making 10s

1 How much taller does each plant need to grow to be 10 cm tall?

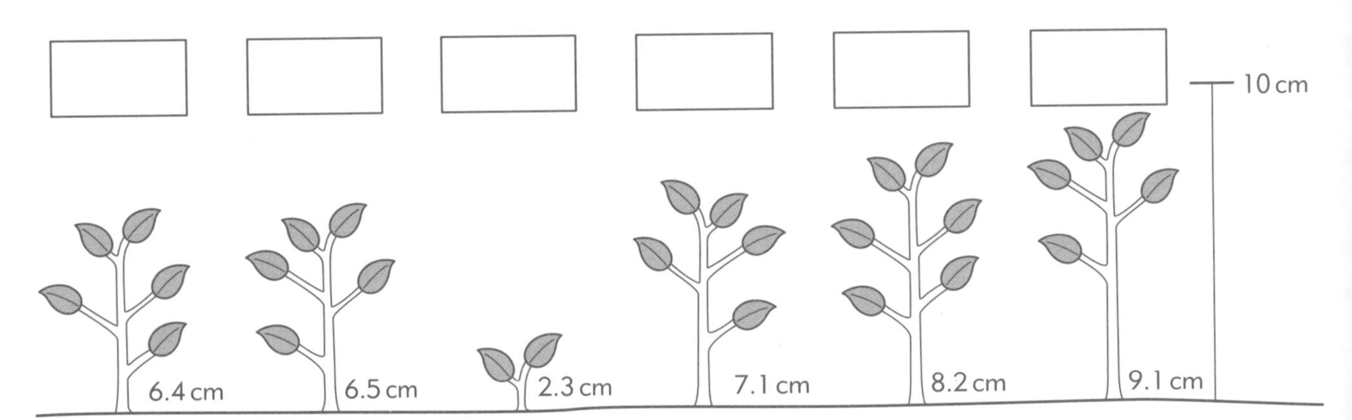

2 Each barrel can hold 10 litres of water. How much more water can each one hold?

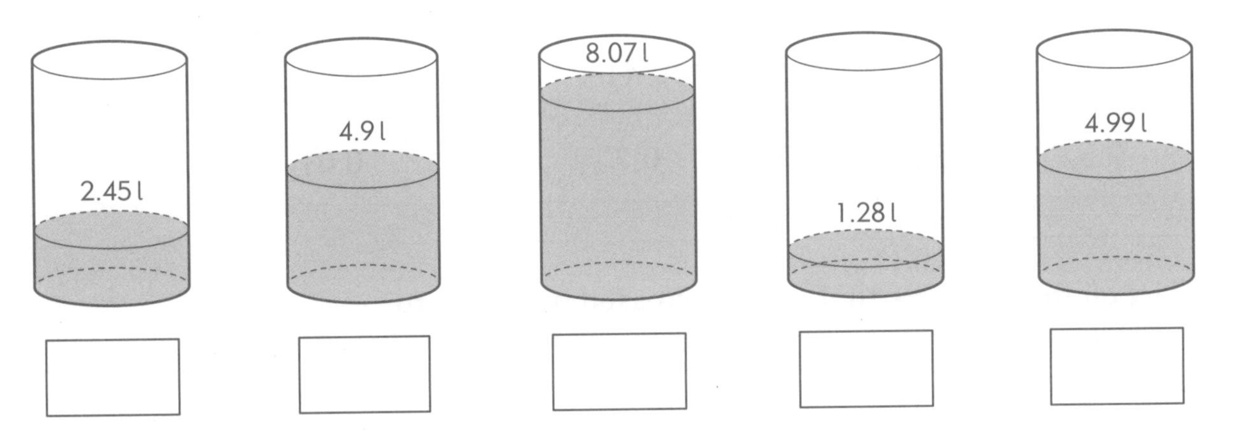

3 Each of these items was bought with a $10 note. How much change did each person get?

see Student Book page 97

Decimal puzzles

In a magic square the sum of each row, each column and each diagonal is the same.

1 **Grid A is a magic square with whole numbers.**

a Work out the missing number.

b Use the numbers in the magic square to make a decimal magic square in grid B. Each decimal should have one decimal place.

c Subtract 5.5 from each number in the decimal magic square. Write the answers in grid C. Is it still a magic square?

Grid A

226	219	224
221		225
222	227	220

Grid B

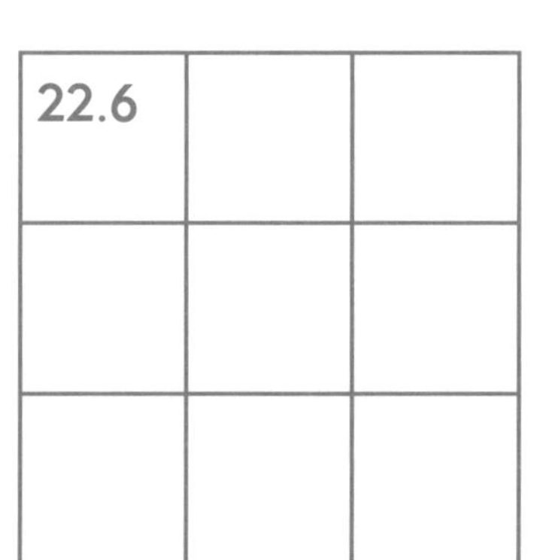

Grid C

2 **Grid D shows another magic square.**

a Find the missing numbers.

b Halve each number and write the answers in Grid E.

c Use the given number as a starting point to make a decimal magic square in Grid F. Each number must have two decimal places. The total of each row, column and diagonal must be 0.42.

Grid D

3.1	12.2	5.7
	7.0	4.4
8.3	1.8	

Grid E

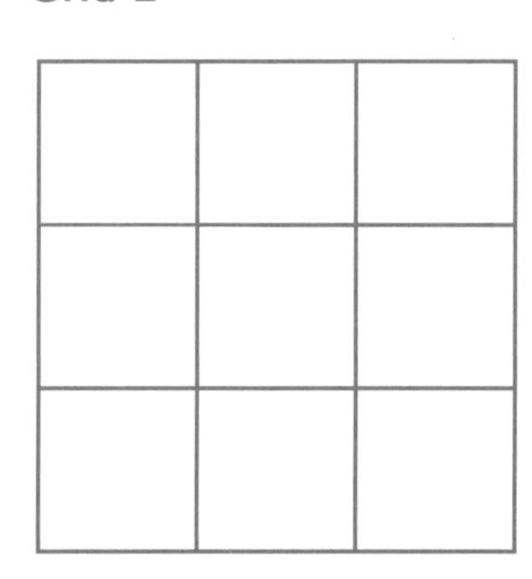

Grid F

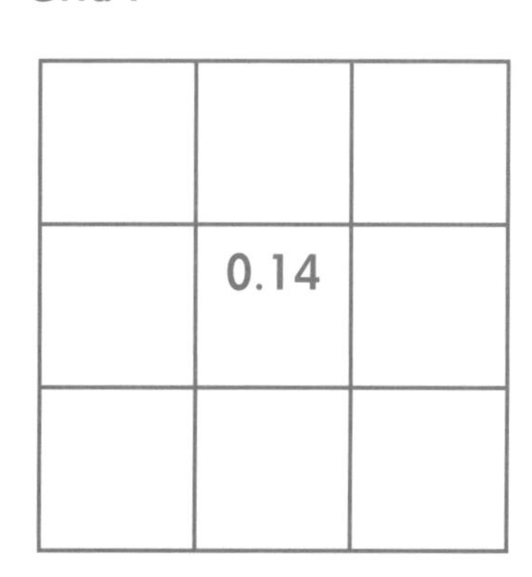

see *Student Book page 99*

Frequency tables

1 Complete the frequency table to show how many of each shape there are in this set.

Shape	Tally	Frequency
Hearts		
Stars		
Rectangles		
Moons		
Circles		
	Total	

2 Use the grouped frequency table to organise this set of test results.

8 7 6 6 6 8 7 5 6 5 4

9 10 10 9 9 7 6 3 8 9 10

3 5 6 7 8 8 8 8 9 4 6

Results	Frequency
1 – 2	
3 – 4	
5 – 6	
7 – 8	
9 – 10	

see Student Book page 101

Bar line graphs

Use these grids to draw the graphs for Student's Book page 102.

Grid A

Grid B

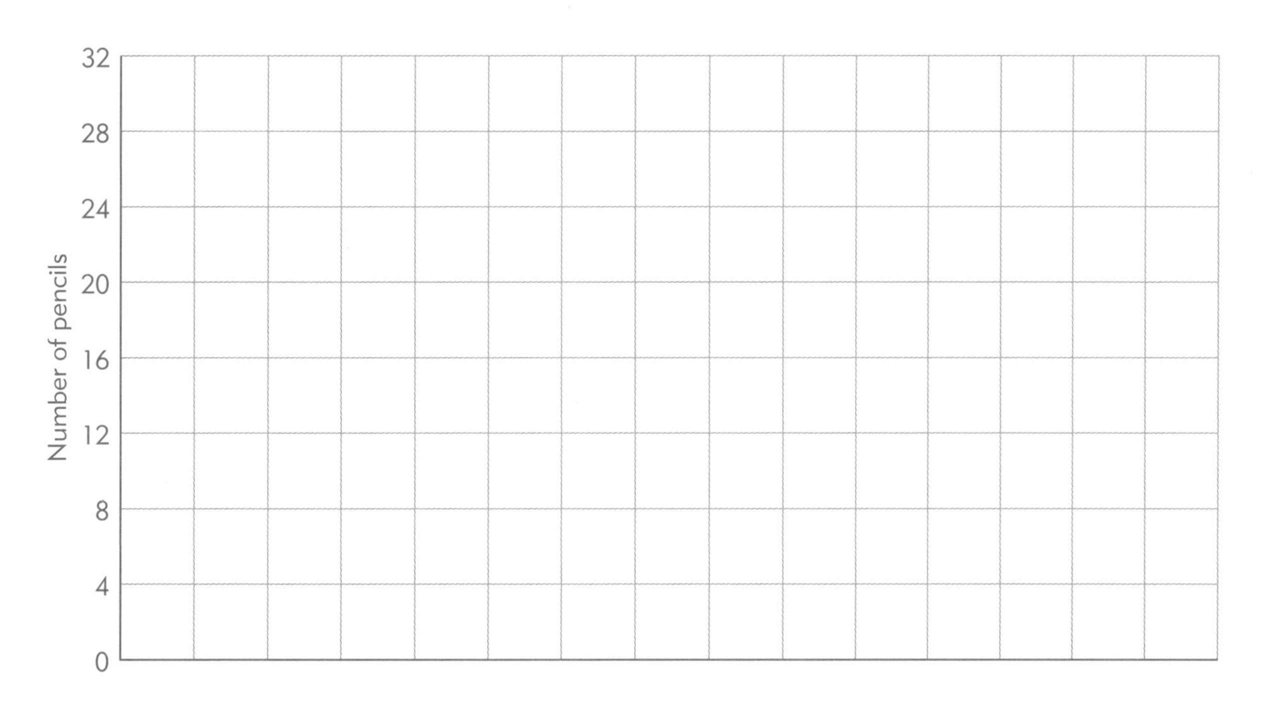

see Student Book page 102

More line graphs

Grid A

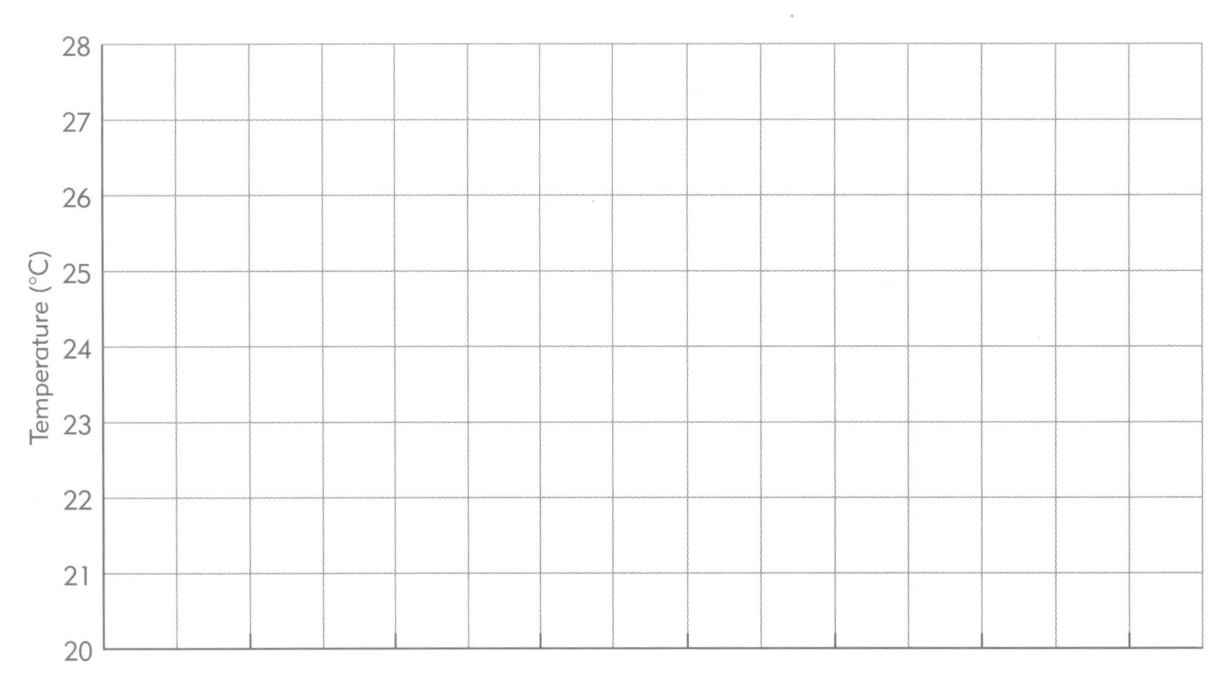

Grid B

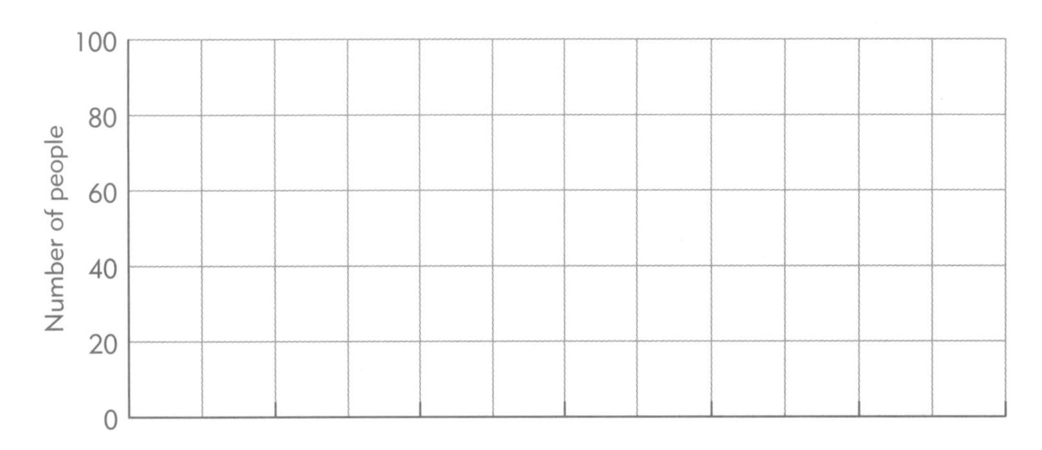

see Student Book page 105

Mental division strategies

1 **Try to calculate these in your head. Check your answers with a calculator.**

	My answer	Calculator answer
600 ÷ 10		
60 ÷ 10		
1900 ÷ 100		
190 ÷ 10		
4000 ÷ 100		
400 ÷ 100		

2 **What happens to the digits of any number divided by the following numbers?**

a 10 ____________________

b 100 ____________________

3 **Now try these:**

a 140 ÷ 20 = ___

b 480 ÷ 60 = ___

c 180 ÷ 30 = ___

d 720 ÷ 90 = ___

4 **Write an instruction for a quick way to divide any number by 30.**

____

____

see Student Book page 108

More division

In this grid the number sentences can be read from left to right **and** from top to bottom.

60	÷	4	=	15
÷		÷		÷
6	÷	2	=	3
=		=		=
10	÷	2	=	☆ 5

Use the numbers in the box to correctly complete each division grid.

1 2 2 5 10 10
10 50 100

☐	÷	☐	=	☐
÷		÷		÷
☐	÷	☐	=	☐
=		=		=
☐	÷	☐	=	☆

2 3 6 8 9
16 24 144

☐	÷	☐	=	☐
÷		÷		÷
☐	÷	☐	=	☐
=		=		=
☐	÷	☐	=	☆

see Student Book page 109

The sawmill

Planks 5 m long are delivered to the saw mill. The workers cut the planks into different lengths.

iStockphoto.com

1 **How long would each piece be in centimetres if they cut a plank into:**

a 4 quarters? _________

b 2 halves? _________

c tenths? _________

d 8 pieces? _________

2 **How many planks would be needed to cut each of the following sets of wood?**

How many pieces of wood are left over as scrap each time? How long are the pieces of scrap wood?

a 6 pieces, each 250 cm long _______________________________

b 9 pieces, each 150 cm long _______________________________

c 5 pieces, each 120 cm long _______________________________

3 **A plank was cut into eight equal pieces. Five pieces were sold. What length of wood was left over?**

____

4 **Eight planks of wood were cut into 50 cm lengths. How many pieces of wood did this make?**

____

see Student Book page 110

What will you do with the remainder?

Solve these problems. Decide whether it is more sensible to write the remainder as a fraction or leave it as a whole remainder.

1 **A carpet layer has a huge roll of stair carpet 250 m long. He cuts it into 9 m lengths.**
He gets __________ pieces.

2 **A light aircraft flies 457 km using 8 barrels of fuel.**
It travels __________ kilometres for each barrel.

3 **Two children are wrapping gifts.**

a They cut 250 cm of ribbon into 10 pieces.
Each piece is __________ long.

b How many 20 cm long ribbons can they cut from a length of 250 cm? __________

4 **$75 is divided equally between 4 people.**
Each person gets __________.

5 **Suggest 3 other ways to divide $75 into equal amounts.**

____

____

____

see Student Book page 112

Shapes and nets

Trace and cut out this net.

Use it to make a cube.

see *Student Book page 115*

Birdwatching

In an hour a birdwatcher saw these birds:

4 pigeons

6 ducks

12 starlings

24 sparrows

2 wild geese

1 owl.

1 **Write down the ratio of:**

a starlings to pigeons ________________

b wild geese to starlings ________________

c ducks to starlings ________________

d sparrows to each of the other birds ________________

In a survey of water birds, these were the ratios:

ducks to geese 6 to 1

ducks to swans 5 to 1

2 **If there were 30 ducks, how many geese and swans were there?**

________________ **geese** ________________ **swans**

3 **If 12 ducks flew away, what would the new ratio be of:**

a ducks to geese? ________________

b ducks to swans? ________________

see Student Book page 118

Colour the correct proportions

Colour each pattern in the given proportions

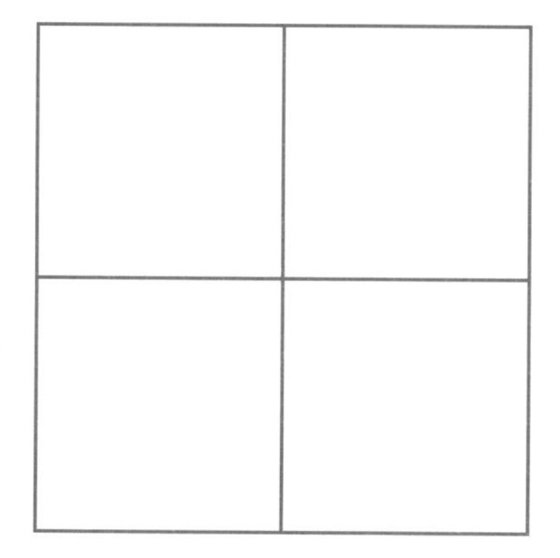

$\frac{1}{2}$ yellow $\frac{1}{2}$ blue

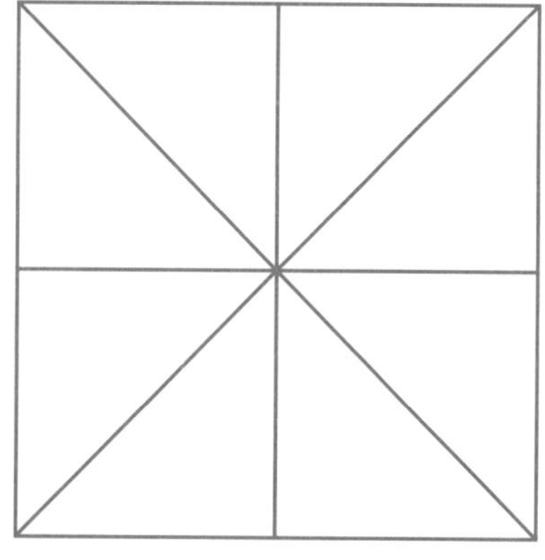

$\frac{1}{2}$ red $\frac{1}{4}$ green
$\frac{1}{4}$ blue

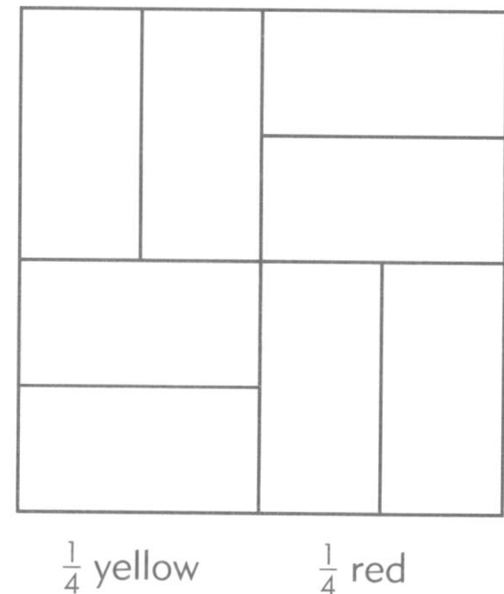

$\frac{1}{4}$ yellow $\frac{1}{4}$ red
$\frac{1}{2}$ blue

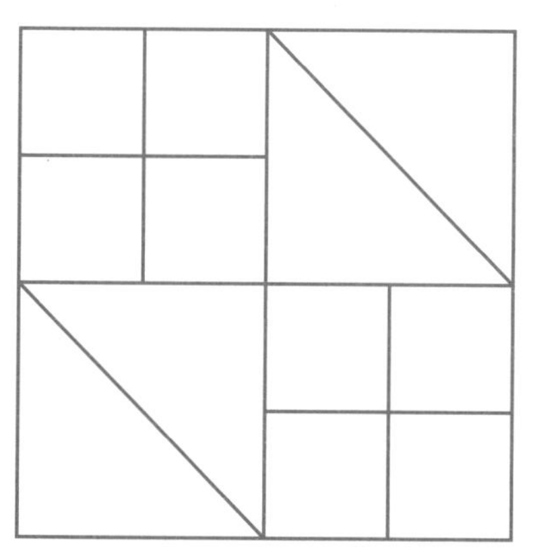

$\frac{1}{4}$ red $\frac{1}{4}$ blue
$\frac{1}{4}$ yellow $\frac{1}{4}$ green

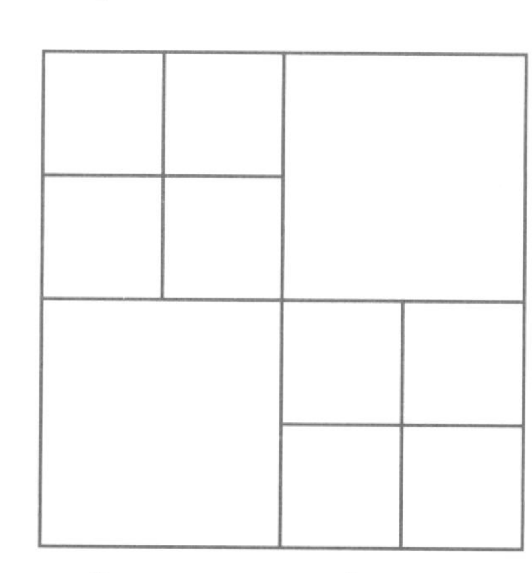

$\frac{3}{4}$ red $\frac{1}{8}$ yellow
$\frac{1}{8}$ blue

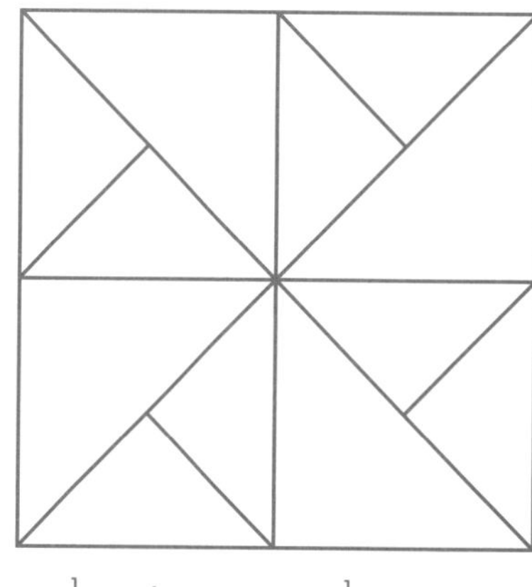

$\frac{1}{4}$ red $\frac{1}{4}$ green
$\frac{1}{2}$ yellow

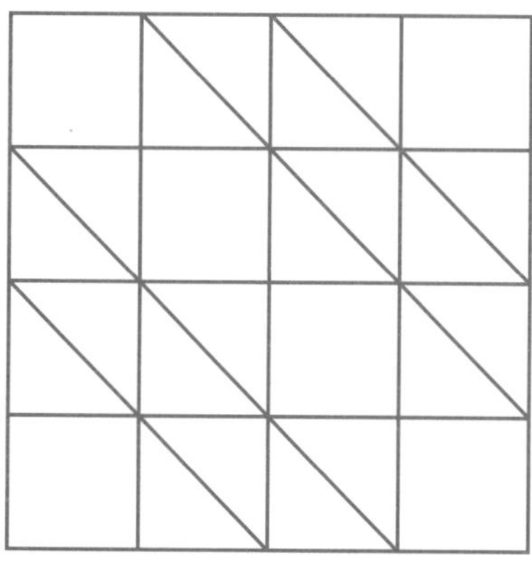

$\frac{1}{4}$ red $\frac{1}{4}$ blue
$\frac{1}{4}$ green $\frac{1}{4}$ yellow

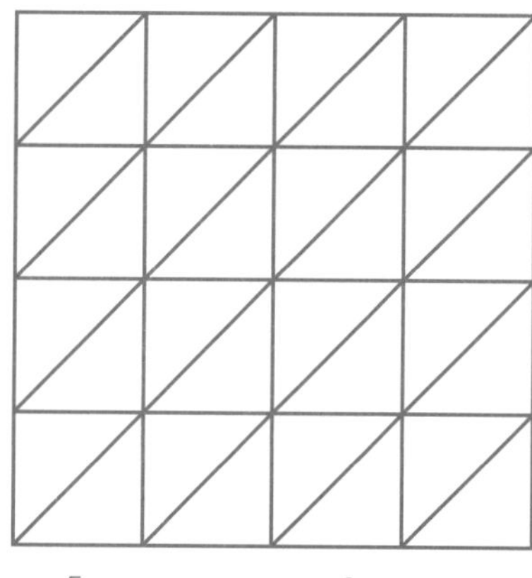

$\frac{5}{16}$ green $\frac{1}{2}$ blue
$\frac{1}{16}$ red

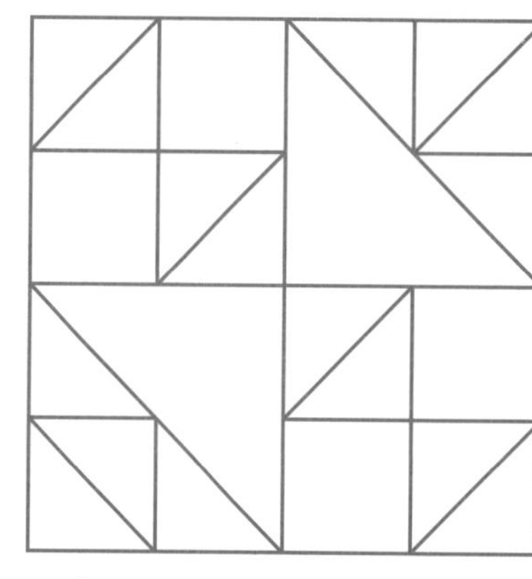

$\frac{1}{4}$ red $\frac{6}{16}$ yellow
$\frac{1}{4}$ green $\frac{5}{8}$ blue

see Student Book page 119

Ratio

1 **List 14 ratios to compare these children.**

For example, boys to girls is 6 to 8.

2 **A supermarket sells both products in special packs:**

2 bottles of shampoo with 1 free bottle of conditioner.

3 bottles of bath foam with 1 free bottle of lotion.

What is the ratio of:

a conditioner bottles to shampoo bottles?

b body lotion bottles to bath foam bottles?

see Student Book page 122

Recipes and proportions

The students at a school want to make chocolate cakes to sell at the school fete.

This is the recipe for making one cake.

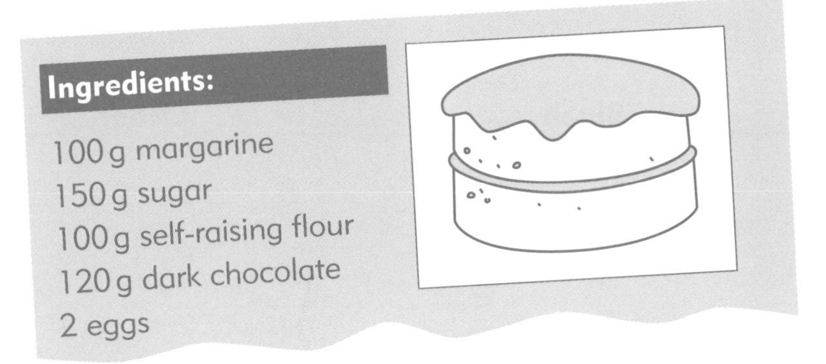

1 A class wants to bake 10 cakes. How much of each ingredient will they need?

__ **g margarine**

__ **g sugar**

__ **g self-raising flour**

__ **g dark chocolate**

__ **eggs**

2 If a class has 12 eggs and as much of the other ingredients as they need, how many cakes can they bake? ____________

3 What mass of sugar will a class need to make 8 cakes? ____________

4 One class sells one cake for $8.75. How much money will they make if they sell:

a 10 cakes ____________

b 8 cakes ____________

c 25 cakes ____________

d 100 cakes? ____________

5 Sanjay's class sold their cakes for $8 each. If they raised $232.00, how many cakes did they sell?

____

see Student Book page 123

Bubble percentages

Shade the bubbles different colours. Then write the percentage of bubbles in each colour.

Red ______________ **Orange** ______________ **Yellow** ______________

Green ______________ **Blue** ______________ **Purple** ______________

Grey ______________ **Black** ______________ **Other** ______________

see Student Book page 124

Problems involving percentages

1 **Shade the correct percentage of each 10 × 10 square.**

a 15%

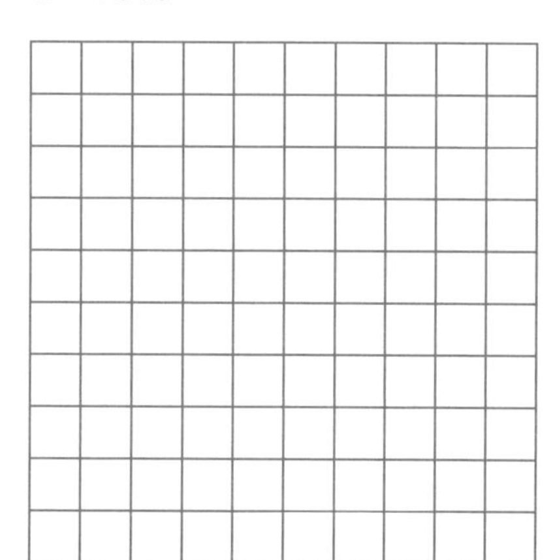

b 20%

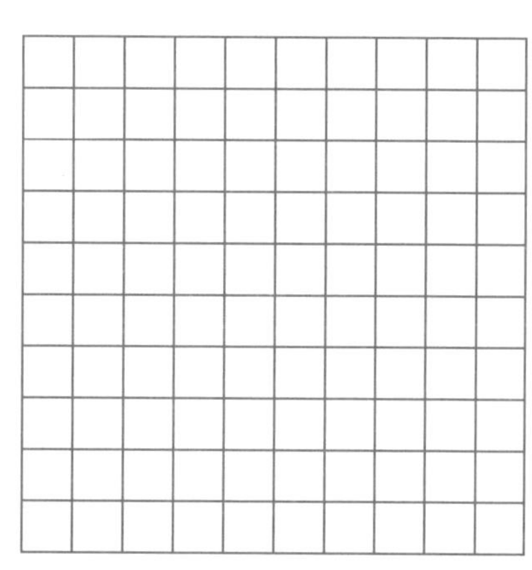

c 1%

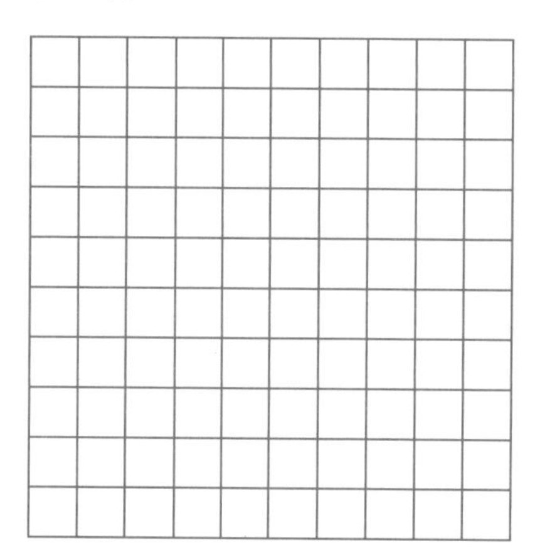

2 **Shade the correct percentage of each shape.**
(Hint: Convert the percentage to a fraction and simplify it!)

a 25%

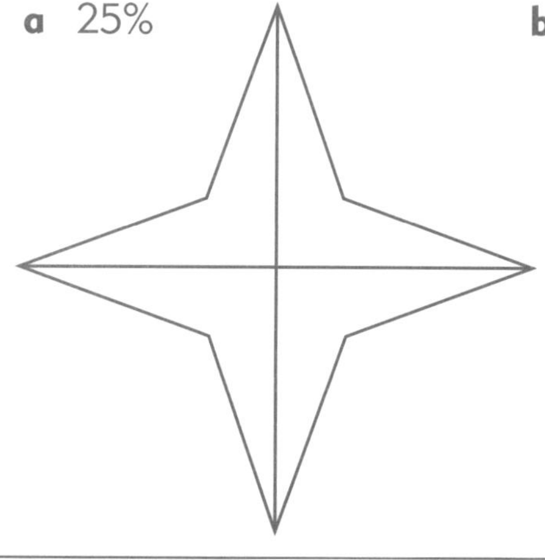

b 40%

c 75%

3 **Complete the sentences.**

a Half is the same as ________%.

b A quarter is the same as ________%.

c One-tenth is the same as ________%.

d Three-fifths is the same as ________%.

4 **Work out these amounts.**

a 100% of $5 is ________________.

b 50% of $5 is ________________.

c 1% of $10 is ________________.

see Student Book page 124

Percentages, decimals and fractions

A number can be written as a fraction, decimal or percentage.

For example, $\frac{1}{2}$ = 0.5 = 50%.

1 **Complete the chart.**

Fraction	Decimal	Percentage
$\frac{1}{2}$	0.5	50%
$\frac{1}{10}$		20%
$\frac{4}{10}$		40%
$\frac{7}{10}$	0.7	
		2%
		17%
		27%
	0.88	
	0.9	
	0.45	
		60%

2 **Show where each of these fractions would fit on the 0–100% line**

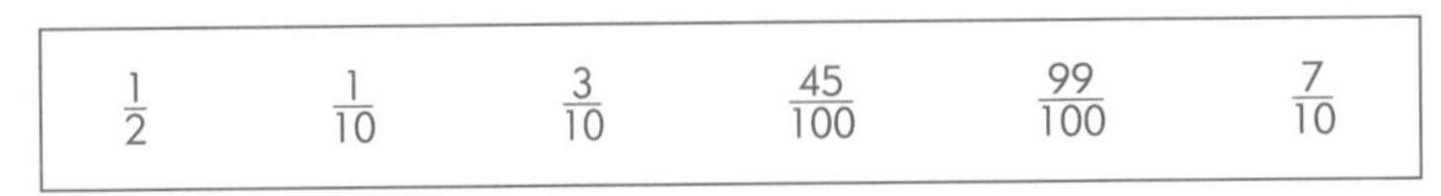

see Student Book page 125

Equivalent fraction wheels

Remember: To convert a fraction to a decimal, you divide the numerator by the denominator.

For example, $\frac{1}{4} = 1 \div 4 = 0.25$

This is a fraction wheel for $\frac{1}{2}$. It shows equivalent fractions and decimals. There are other fractions which could go on this wheel, for example, $\frac{10}{20}$, $\frac{8}{16}$ or $\frac{50}{100}$.

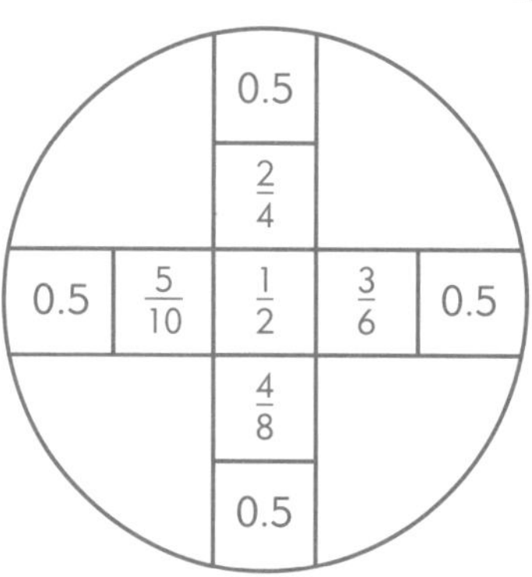

Complete these equivalent fraction wheels.

1 **a**

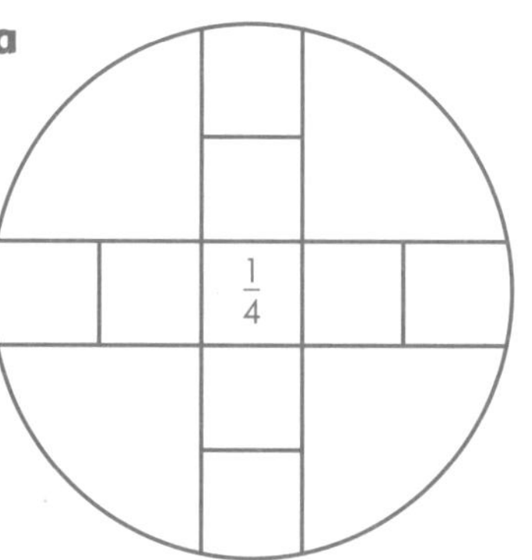

b

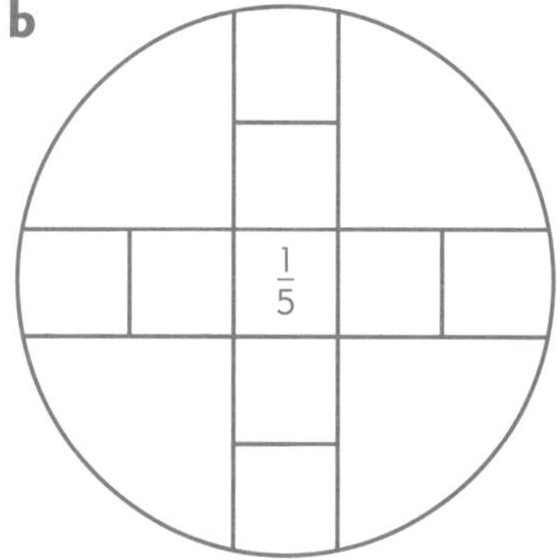

c

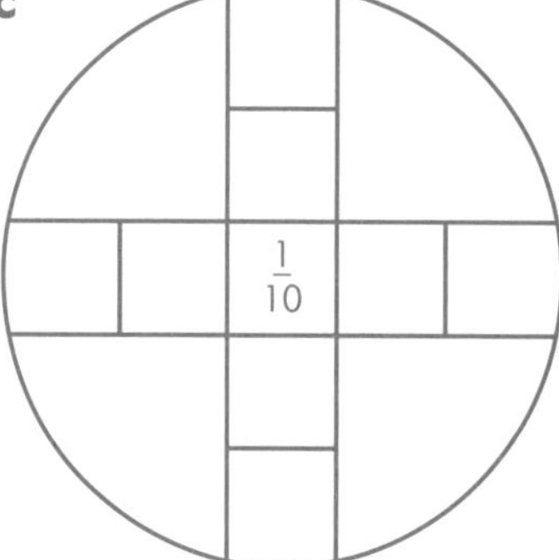

2 **a**

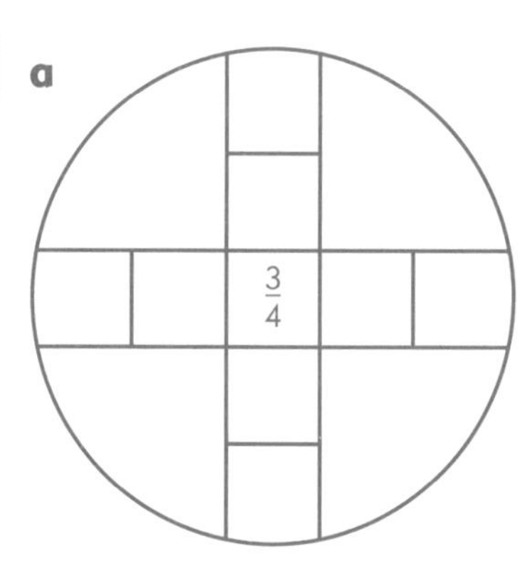

b

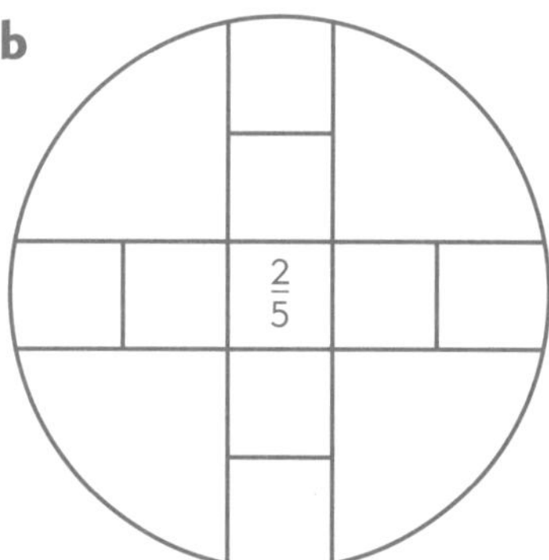

c

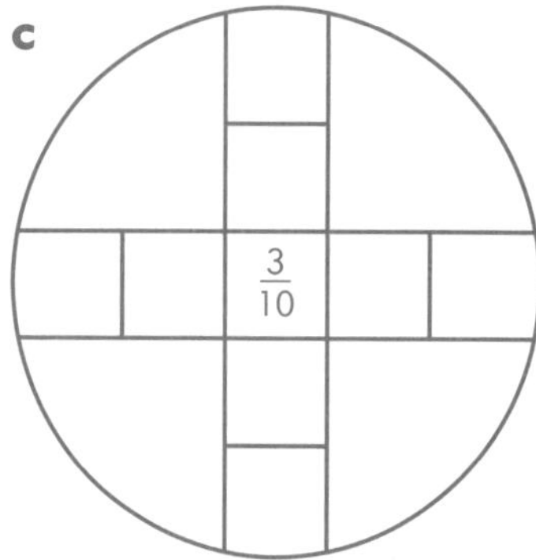

see Student Book page 128

Ordering equivalent fractions

1 Complete this table.

					$\frac{1}{2}$					
Tenths	$\frac{1}{10}$				$\frac{5}{10}$	$\frac{6}{10}$				
Decimals		0.2		0.4				0.8	0.9	
Percentages			30%				70%			100%

2 Arrange each set of fractions in order from smallest to greatest.

a 0.25 $\frac{3}{4}$ $\frac{2}{10}$ $\frac{0}{01}$ 0.7 50%

b $\frac{9}{10}$ 75% 0.8 $\frac{1}{2}$ 25% $\frac{2}{5}$

c 60% 0.66 $\frac{6}{100}$ $\frac{1}{2}$ 0.4

3 Write down two numbers from each set above that have a sum of 1.

Set A: ☐ + ☐ = 1 Set B: ☐ + ☐ = 1 Set C: ☐ + ☐ = 1

4

a Write down the highest and lowest value in each set.

Set A

Set B

Set C

b Calculate the difference between the two values.

c Show the differences on this number line.

see Student Book page 128

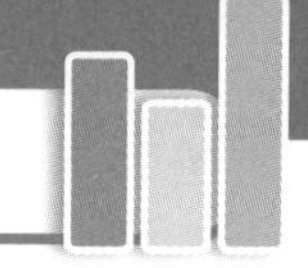

Likely and unlikely events

Draw lines to match the events to the correct place on the scale.

Certain
Likely
Even chance
Unlikely
Impossible

see *Student Book page 130*

Choose your method

1 **Here are some multiplications. Try and solve them without a calculator. Choose your own method.**

a 4 × 239 = ______

b 332 × 7 = ______

c 155 × 9 = ______

d 469 × 6 = ______

e 497 × 4 = ______

f 199 × 9 = ______

2 **Mr Musa's electric shop orders some electronic equipment. Work out the total cost of each order.**

a 9 TV sets at $114 each.

b 7 camcorders at $799 each.

c 925 washers at 9c each.

d 8 extension cables at $8.99 each.

see Student Book page 133

Find the area

Work out the area of each field. Do your working inside the fields.

Remember Area = length × breadth

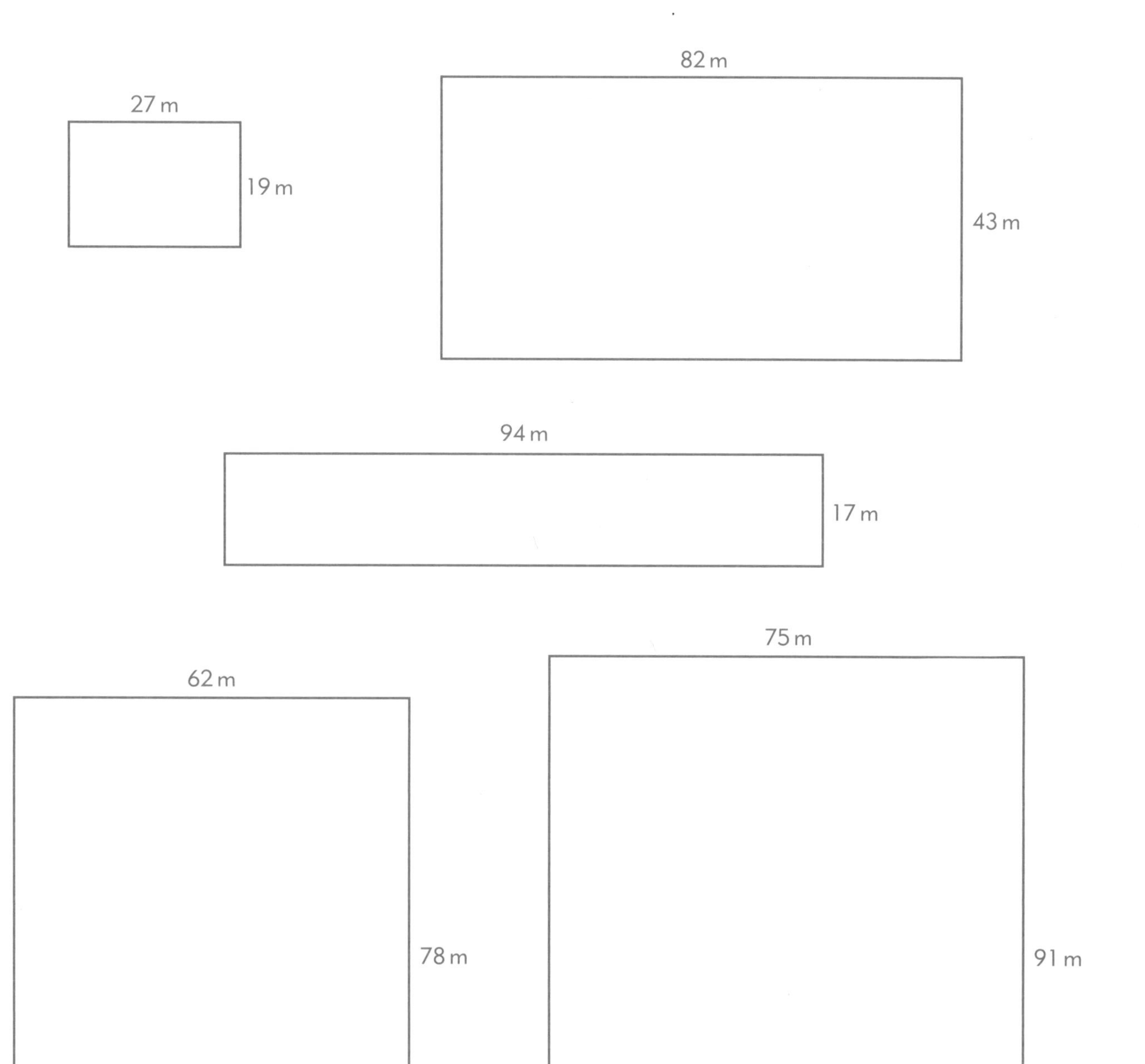

see Student Book page 134

Multiplying decimals

Complete this multiplication table.

×	2	3	4	5	6	7	8	9
1.1								
1.8								
1.7								
2.3								
2.5								
2.9								
3.0								
3.1								
3.3								
3.7								
4.5								
4.8								
5.9								
7.2								
9.6								
9.9								

see Student Book page 137

Work out the mass

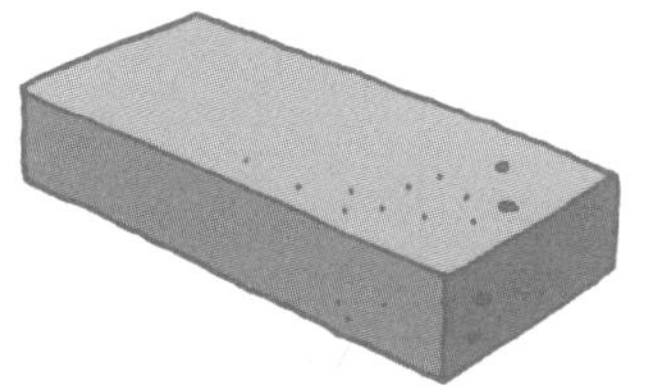

block 3.2 kg

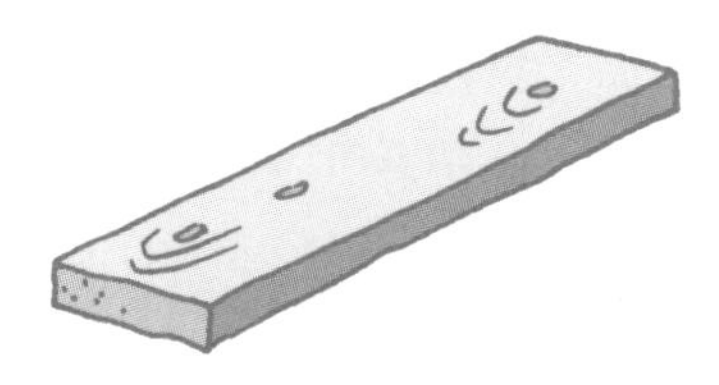

plank 2.8 kg

bag of nails 1.1 kg

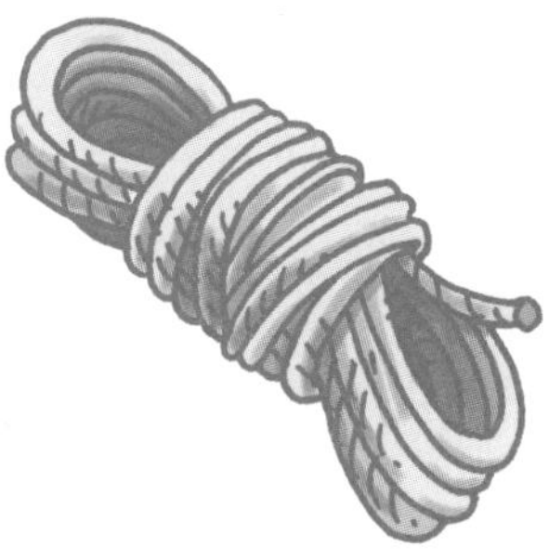

coil of rope 6.4 kg

The mass of each item is given.

Work out the mass of each pile.

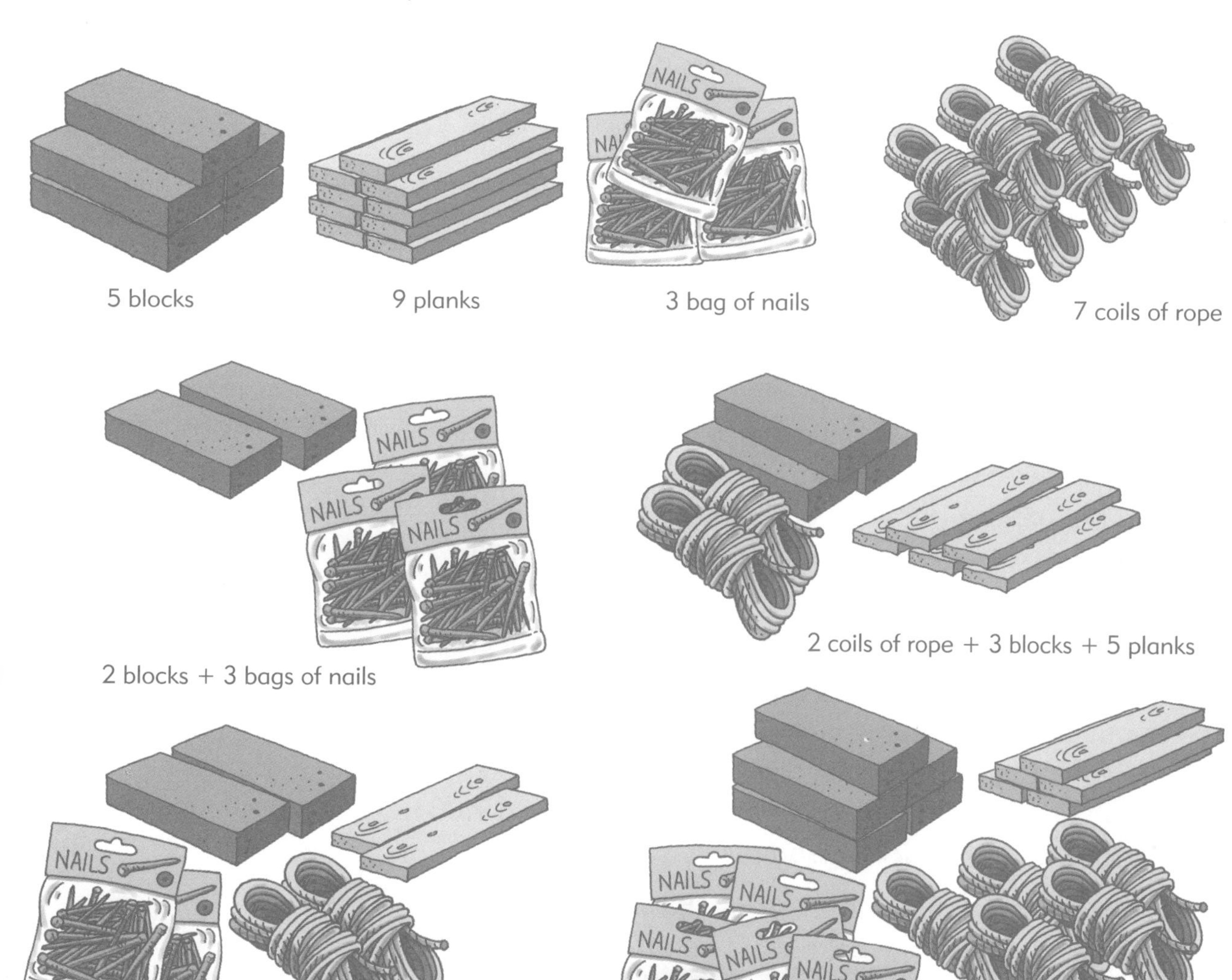

5 blocks

9 planks

3 bag of nails

7 coils of rope

2 blocks + 3 bags of nails

2 coils of rope + 3 blocks + 5 planks

2 of each item

5 of each item

see *Student Book page 138*

Mixed calculations

1 Complete these mixed operations.

> **Remember BODMAS, to give you the correct order of operations:**
>
> **B**rackets
> **O**f
> **D**ivision
> **M**ultiplication
> **A**ddition
> **S**ubtraction

a $\frac{1}{3}$ of $(200 - 50)$

b $(100 + 44) \div 12$

c $(10 + 4 \times 4) + (17 + 3 \times 8)$

d $18 \div 3 + 8 \times 2 + 25 \div 5$

e $\frac{1}{2}$ of $18 + 19 + 20 + 21$

f $\frac{1}{2}$ of $22 - \frac{1}{2}$ of 16

g ($\frac{3}{4}$ of $66) + 55 \div 11$

h $43 + 19 + 15 + 200 \div 2 \times 5$

i $10 \times (16 - 6)$

j $4 \times 4 + 4 \times 4 + 5 \times 5$

k $\frac{1}{2}$ of $50 + \frac{3}{4}$ of 100

l $7 \times 30 + 2 \times 40$

m $\frac{3}{4}$ of $16 + \frac{1}{4}$ of 16

n $(10 + 10 + 10 + 10 + 10 + 10) \div 5$

---

2 Complete these problems. You will first need to write the number sentence, then find the answer.

a Jenny organises a cherry-picking expedition and picnic. The following people confirm they will join her: Mary and her two sisters, James and his four cousins, Leah and her brother and mother. Jenny needs to take one picnic basket for every two people. How many picnic baskets must she take?

b Three friends go out for a meal. They add up the items on the bill and then split the total evenly between them. The items on the bill are as follows: two starters at $6.45 each, three main dishes which cost $8.99, $7.35 and $10.50, and one dessert which costs $9.00. How much must each person pay?

c Class A has 32 students and Class B has 28 students. $\frac{3}{4}$ of the students from each class go on the school outing. How many students go altogether?

see Student Book page 141